Ores to Metals

Ores to Metals

The Rocky Mountain Smelting Industry

James E. Fell, Jr.

293373
University of Nebraska Press ● Lincoln and London

The publication of this book was assisted by a grant from The Andrew W. Mellon Foundation.

Library of Congress Cataloging in Publication Data

Fell, James E   1944–
  Ores to metals.

  Bibliography: p. 311
  Includes index.
  1. Mineral industries—Rocky Mountain Region—History.
2. Smelting—History. 3. Rocky Mountain Region—Industries—History.
I. Title.
HD9506.R582F44   338.4'7'66902820978   79–9093
ISBN 0–8032–1951–2

For My Mother and Father

Contents

Illustrations

Preface

THE MOUNTAINS AND DESERTS OF THE WEST STILL HOLD THE picturesque ruins of abandoned mines. The boarded tunnels, rough-hewn buildings, immense piles of yellow tailings, and narrow headframes that reach upward to the sky stand as silent witnesses to the vitality of men who once wrested minerals from the earth. But there are few ruins to remind a passerby of the smelters that bought the ores, reduced them to bullion and matte, and then drew out the gold, silver, copper, lead, and other metals. Here you might see bushes growing through a stone foundation; there you might note a mound of black slag. Here you might glimpse a small chimney rising amid the aspen; there you might notice a towering smokestack still piercing the western sky. But little else remains. The wreckers left few reminders of this industry that rose, flourished, and died in symbiosis with the mines.

Despite the paucity of ruins, the smelters played an indispensable part in the minerals industry, and their evolution displayed all the main characteristics of American business in the Gilded Age. They appeared in response to the miners' need for a technology that could recover gold, silver, and other metals from ores resistant to better-known methods of reduction. Businessmen tried many processes—some invented locally, others devised in the East, and most of which failed. But the shrewdest, most knowledgeable entrepreneurs adapted techniques long used in Europe and built the first plants largely with eastern capital. Once begun, the industry evolved from one composed of perhaps a thousand small enterprises working ores in isolated mining camps to one composed of several large, integrated firms operating plants in major urban centers, from which they tapped ore markets hundreds and sometimes thousands of miles away. Combinations made the large firms part of even larger enterprises, until nearly the entire industry was absorbed by one giant holding company in the course of the great merger movement that engulfed American business in the late nineteenth and early twentieth centuries.

Because the mines of Colorado lay in the heart of the great western mineral empire, it was no coincidence that the smelters built in the Rocky Mountains formed the center of the industry. The first plants appeared in the midst of the crisis that devastated the mining business during the 1860s. They smelted ores drawn from local mines. Later firms served more distant mining districts. As large interregional ore shipments became possible, the industry concentrated its work in four cities—Denver, Pueblo, Leadville, and Durango. From here the smelters tapped mines as far north as Canada, as far west as California, and as far south as Mexico. A trend toward mergers developed within the industry, and once underway, consolidation continued until nearly all the independent enterprises were absorbed by the American Smelting and Refining Company. At the time of its inception this firm had one half of its reduction capacity in Colorado and drew its primary management from men connected with the industry there. Yet by World War I the collapse of the traditional mining industry had a devastating affect on ASARCO. The company survived, but its plants in the region did not.

Despite the critical part the smelters played in the minerals industry, the number of books and articles written about the production of ores overwhelms the number written about the reduction of ores to metals. An oversight, perhaps, but the mines vastly outnumbered the smelters and had a glamor, an aura of wealth, that gleamed in its own time and continues through the years. Yet ore reduction formed an integral, if less glamorous, part of the minerals industry. As in mining, the production of metal demanded the steady application of capital, labor, and technology. Fortunes were won and lost, some men converted success in business to success in politics, and many made important contributions to the advance of science and metallurgy. This was an industry that has long needed a history.

Like every author, I have incurred a large, perhaps unpayable debt to many people who helped me at one stage or another in research and writing. I would like to cite everyone here by name, but to do so would create a very long list, much to the dismay of anyone inadvertently forgotten and more to the dismay of the publisher. So I wish to offer collective thanks to everyone who helped make the book possible. Yet I do wish to offer a special acknowledgment to the University of Colorado for a research fellowship in 1974 and to Harvard University Graduate School of Business Administration for a Kress fellowship the same year. I also wish to thank the University of Arizona for permission to quote passages from "Nathaniel P. Hill: A Scientist-Entrepreneur in Colorado," which appeared in the Winter 1973 issue of *Arizona and the West*. I also wish to express my appreciation to the editor of the *Business History Review* for permission to quote some materials included in "Rockefeller's Right-hand Man: Frederick T. Gates and the Northwestern Mining Investments," published in the Winter 1978 issue of that journal. Last but not least, I would like to offer a note of thanks to Sue Mills and Janice O'Reilly, who slaved over the manuscript for so many hours.

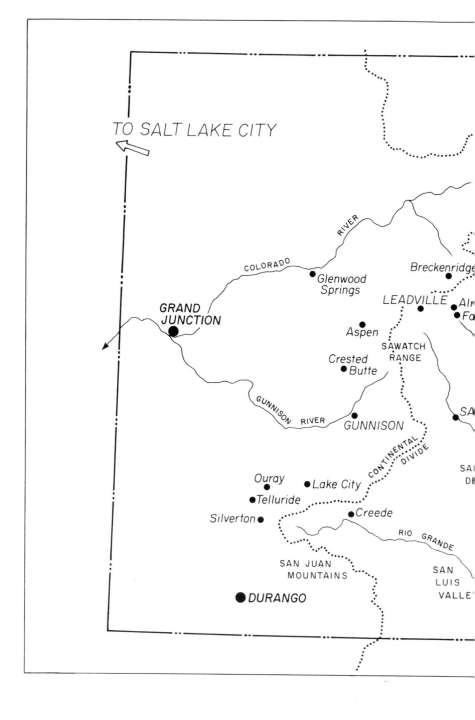

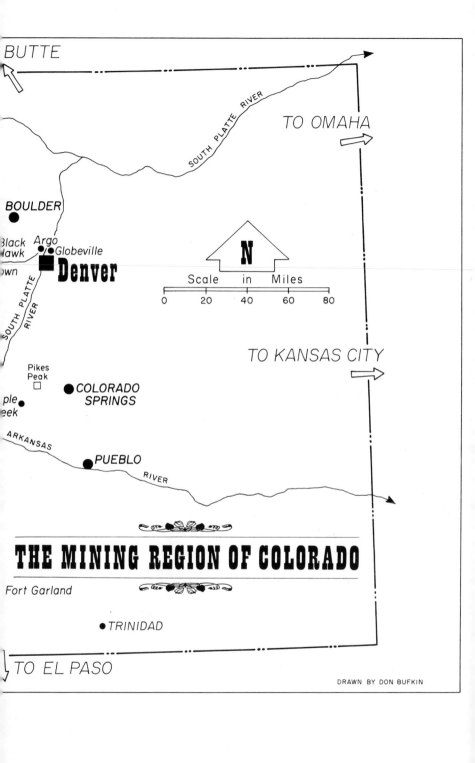

THE MINING REGION OF COLORADO

Fort Garland

●TRINIDAD

↓TO EL PASO

DRAWN BY DON BUFKIN

Ores to Metals

Chapter 1

"A Savan among Us"

Rumors of gold and silver in the Rocky Mountains had drifted down through the years since Spanish times. Rivera, Purcell, Frémont, and others all claimed they had found precious metals in the high country, but their reports were conjectural. No one had ever confirmed them. The California gold rush gave rise to a new spate of rumors; yet not until the winter of 1857 descended on Auraria, Georgia, did anyone make a systematic effort to learn the truth.

It was in that season that William G. Russell took a notion to explore the foothills of the Rockies near the point where the South Platte River enters the high plains on its journey to the sea. Not much is known about Russell, but he had a forceful personality matched by a moustache and beard that made him stand out like a musketeer from a Dumas novel. Russell had been in the Rockies before. In 1850 he had driven horses to California, and he may have found a few flakes of placer gold while passing through the mountains.

Early in 1858 Russell set out from Georgia with his brothers and a few friends. At Leavenworth in Kansas Territory they joined a band of Cherokees led by John Beck, who, like Russell,

1

may have found gold in the high country while on his way to California in 1849. With their ranks now swelled to more than a hundred, the adventurers crossed the prairie to reach the South Platte, then traced the waterway to its rendezvous with a smaller stream known as Cherry Creek. After pitching their tattered canvas tents, Russell's men fell to work on the sands and gravels. Here on the future site of Denver they found tiny particles of placer gold, confirming the rumors that had abounded for so long.

Russell's men were exultant, but their dreams of instantaneous wealth gave way to frustration and disappointment. "Pay" from the "diggins" was meager, so meager it hardly justified the backbreaking toil of shoveling and panning, the crude techniques of placer mining. Nearly all the adventurers became discouraged and abandoned the search that fall, but by chance an itinerant trader bound for Missouri happened upon the camp. He traded for a few flakes of gold and went on his way. Once he reached the first line of settlements farther east, his exaggerated account of riches spread from town to town like a prairie fire. Merchants and newspapermen boomed the discovery to bonanza proportions, touching off the Pike's Peak gold rush that came the next year. But the Cherry Creek placers hardly justified the coming frenzy, and a fiasco loomed as winter settled over the high plains.[1]

Even before the end of 1858, the first treasure seekers arrived on Cherry Creek. It was not long before they realized that the stories of easy wealth had been vastly exaggerated. Many cursed the Pike's Peak "humbug," turned around, and went home; but others were not so easily discouraged. Miners experienced in Georgia and California perceived that the gravels at the base of the Rockies contained flakes of gold washed down from the white-tipped mountains that rimmed the western skyline. And so, as the year of discovery drew to a close, bands of prospectors set out to ascend the partly frozen streams that flowed out of the Front Range into the South Platte.[2]

In one of those companies was John H. Gregory, an old Georgia miner who arrived in the foothills late in the year. He and several companions decided to trace into the mountains a

stream known as the Vasquez Fork of the South Platte, a name
soon changed to Clear Creek as the American tide overwhelmed the Spanish influence remaining in that part of the
high country. Battling snow and ice and cold, the adventurers
struggled up the narrow, high-walled canyon until the stream
branched at the base of a steep ridge. Choosing the right-hand
fork, soon known as North Clear Creek, they pushed on to
where the narrows opened into a series of high hills timbered
with firs. There in January 1859 John Gregory found flakes of
placer gold in what became the town of Black Hawk.

Gregory and his companions, however, wanted to locate the
source of the placer—and so they kept looking. They nearly
perished in a late winter snowstorm, but on May 6 their
persistence was rewarded when Gregory uncovered the rust-
colored outcropping of the lode that bears his name. This was
the first vein of gold discovered in Colorado. Like prospectors
everywhere, Gregory and his friends tried to keep their good
fortune secret, but this proved impossible. The news carried
down to embryonic Denver City, and up Clear Creek came a
rapid influx of fifty-niners. Some concentrated on placering at
the Gregory diggings, but others spread out into the surrounding hills, where they located the Bobtail, Bates, Gunnell, Illinois, and other lodes. Before long the ragged tents and hastily
improvised log cabins thrown up along the banks of North
Clear Creek grew into the towns of Black Hawk, Central City,
and Nevadaville.[3]

While Gregory and his companions were trudging upstream
to the future site of Black Hawk, George A. Jackson, a veteran
California miner, was tracing another path to gold. He ascended the south fork of Clear Creek until he came to a stand of
barren willow trees hard by a smaller current that flowed into
the main stream. Here at the confluence he panned telltale
flakes of placer gold. Unlike Gregory and his friends, Jackson
managed to conceal his discovery for some time, but the news
leaked out a few months later, and Pike's Peakers rushed to
Jackson's "diggins." By summer fortune hunters were sluicing
the sands and gravels all the way from the forks of Clear Creek
to the site of Jackson's discovery, soon to be the town of Idaho

Springs. Other fifty-niners pushed farther west, where they located placer and vein gold that gave rise to the mining camps of Spanish Bar, Empire, Elizabethtown, and Georgetown.

Prospectors poured into the high country all year long, but not all sought their El Dorado in the sands and gravels of Clear Creek. Some traced the tawny-colored foothills to a stand of rotting cottonwoods on South Boulder Creek, about twenty-five miles northwest of Denver City. Here they located a placer they called the "deadwood diggins," which prompted more extensive mining in the mountains to the west. Still other fifty-niners crossed the Front Range of the Rockies into the great valley of South Park, traced meandering streams across to the northwest rim, and found placer and lode gold near what became the mining camps of Fairplay, Tarryall, and Buckskin Joe. And well to the south some prospectors hurrying west along the Arkansas River followed that stream into the central Rockies rather than heading to the popular tributaries of the South Platte. They realized their dreams of finding placer gold at Kelly's Bar, Cache Creek, and other sites now long forgotten.[4]

No one knows now many people came to the mountains in 1859. Some estimated the number at 100,000, but this seems a vast exaggeration—a figure inflated by enthusiasts to promote the country. A better estimate would be fewer than 25,000. Two years later the first official census showed the population to be about 23,000. In contrast, something like 80,000 newcomers poured into California in 1849. But, regardless of the number of people who arrived in the Pike's Peak Rush, there was chaos throughout the region. Denver City, Central City, and other "cities" were little more than a confusion of tents, shacks, and log cabins. And not until the middle of the secession crisis of 1861 would Congress create Colorado Territory. In the meantime, what "legal system" prevailed was hardly more than an arcane jumble of local rules and familiar customs enforced by popular opinion or lynch law. The pioneers tried to re-create the social, economic, and political institutions they had known, but their efforts only led to a mosaic that differed from camp to camp and town to town.

Despite the chaos, the first miners shaped the outline of the minerals industry for the next twenty years. The heart of all their activity would lie on the forks of Clear Creek in the country opened by the discoveries of Gregory and Jackson. The most important towns would be Black Hawk, Central City, and Nevadaville, which the first territorial legislature grouped as Gilpin County. A close second would be Empire, Georgetown, and Idaho Springs, which territorial lawmakers included in Clear Creek County. The amount of gold and silver mined in those two regions made up nearly two-thirds of Colorado's production until the epochal year of 1879.[5]

Like other western mining regions, Colorado was part of cordilleran America—the great highland formed by the Rocky Mountains on the east, the Sierra Nevada and the Cascade range on the Pacific slope, and the arid plateaus in between. When the mountains were created centuries ago, hot mineral-bearing liquids flowed upward toward the surface of the earth. Sometimes they pressed into fissures and cracks in the crust, leaving veins of metal. At other times the molten matter dissolved the existing rock and substituted other materials, creating replacement deposits. Whether the minerals formed a true vein or a replacement deposit, the result was the same—a sheet of solid rock in which the accumulation of gold, silver, or other metals varied in length, depth, and thickness. It was this crazy-quilt pattern that accounted for the high risk and heartbreaking failure that characterized the mining industry.

In Colorado the minerals lay in several forms. Some were pyrites—compounds of iron, copper, and sulfur. Others were galena-sphalerites—mixtures of lead, zinc, and sulfur. Still others were combinations of the first two—compounds of iron, copper, lead, zinc, and sulfur. And a fourth group, unimportant to the early miners, were tellurides—compounds of the metal tellurium. Encased within the minerals were gold as the free element and silver as silver chloride or silver bromide. Such deposits held valuable metals besides gold and silver, but the ores were complex, and except for placers they would be difficult to process—something no one realized in the optimistic days of 1859.[6]

In the millennia that followed deposition, the incessant activity of the atmosphere created a three-level system of ores. As air, water, and sun oxidized and disintegrated the surface minerals, rainwater and melting snow dissolved the silver compounds and washed away some gravel, leaving an outer deposit richer in gold, copper, lead, and zinc. (Running water also carried some of the gold downstream, forming the placers found by Russell, Jackson, and others.) Below the surface lay a second layer known as gossan. Here the elements oxidized and enriched the minerals down to depths of fifty or even a hundred feet, forming an ore system that could be easily mined and processed. Below this region, however, was a vast third tier of sulfides that proved far less rich and far more difficult to work than the ores above.

In 1859 fortune hunters like Gregory and Jackson sought their winnings from "dirt" with pans, cradles, and sluices—the ageless techniques of placer miners. These simple devices were all similar in principle. Each depended upon the high specific gravity of gold—its great weight relative to that of other substances—to cause it to be left behind when a current of water washed away other substances *presumed* to be worthless.

Once subterranean mining began, Colorado's miners had to employ new techniques to crush the gossan and free the gold. At first they turned to a crude device long employed by the Spaniards—the arrastra. This consisted of a circular stone trough and several heavy stones known as "mullers" that hung from an arm attached to a central pivot. The miners placed their ore in the trough, then oxen, mules, or other draft animals dragged the "mullers" over it to crush it and free the gold, which was later recovered by sluicing. The first arrastras appeared in July 1859, only two months after Gregory's discovery. By the end of the year several were in operation along the banks of North Clear Creek, each earning as much as $200 daily. Arrastras were easy to finance and simple to construct, but they were slow and inefficient—distressing faults to men bent upon instant wealth. Throughout the Rockies they were little more than transitory devices that soon gave way to stamp milling.[7]

Stamp mills were far more characteristic of an industrial society. Stamps were heavy iron blocks attached to wooden or iron rods that rose and fell in accord with a revolving horizontal beam. With monotonous, incessant crashes, they reduced hard rock to sand, which was then washed by a stream of water over large copper plates impregnated with mercury. This substance formed an amalgam—a kind of alloy—with the gold. Later, mill operators heated the amalgam in a retort. This broke down the alloy, vaporized the mercury (which was condensed and reused), and left behind the gold, which could be cast into bars. The principle was simple, but stamp milling required a knowledge of engineering, a supply of semiskilled labor, and access to capital. The appearance of mills marked a step in the passing of the fortune hunters.[8]

Stamp milling developed rapidly in 1859—so rapidly that the essential machinery must have been on its way to Colorado before Gregory made his famous discovery on May 6. Though they required capital, mills were relatively inexpensive to build and operate, and the first plants came into production during the summer. Returns were often large, some enterprises recovering as much as $400 in gold a day. Such spectacular yields, however, created an unjustified optimism that spurred mining companies and independent milling outfits into overconstruction. Only rich lodes, like the Gregory and the Bobtail, could produce large supplies of ore worth more than the $30 a ton, which enabled both mine and mill to work profitably. Since many ores paid no more than the cost of milling, idle stamps quickly become commonplace.[9]

Yet the boom was on. Good times brought in settlers, stimulated capital investment, and provided a measure of prosperity. The tents, shacks, and log cabins of the fifty-niners gave way to more substantial buildings, many with tall, narrow windows and false fronts. The bustle of life in the narrow streets was punctuated by the thud of underground explosions, the rattle of steam engines, the shriek of whistles, and the crash of stamps that echoed across the canyons. Trees disappeared to provide fuel for the furnaces and timber for the mines. And with all this activity the yield of gold rose steadily from just about nothing in 1858 to perhaps $3,000,000 in 1861 and about $4,500,000 in 1863. Many back East looked to Colorado as a land where they might improve their lot in life, if not get rich quick.[10]

Yet this was an illusion. By the end of 1863 the miners had exhausted the best placers and the easily processed supplies of gossan. In the mines, all that remained at levels no more than one hundred feet below the surface was a vast tonnage of un-oxidized ores known as sulfurets because they smelled so strongly of sulfur. They proved resistant to stamp milling be-cause they held the gold in the form of a solid solution. The metal was so finely divided that crushing freed little of it. But no one understood this. When mineowners sent ores to be mil-led, they learned to their horror that the recovery rate dropped from as much as three-quarters of the assay value, the normal yield, to less than one-quarter of what they expected. And sometimes there was no yield at all. A few people had noticed the trouble as early as 1860, but in those days miners had dismissed the problem as an aberration peculiar to certain lodes. By the end of 1863, however, the impasse had become widespread. More than anything else, the inability of the stamp mills to recover gold from sulfurets brought on the se-vere economic contraction that convulsed the high country over the next few years.

Yet the failure of the stamp mills was only one problem facing the industry. As mineowners blasted farther into the earth, they had to raise more capital to buy new machinery and replace flimsy timbering. The need for money happened to coincide with the surging inflation created by the Civil War, an inflation that caused many investors to believe that shares in a gold mine were a better means of preserving wealth than the depreciating paper money. Taking advantage of this senti-ment, many mineowners floated stock issues that touched off a speculative boom in Boston, New York, and other cities. Some offers were legitimate, but others were sheer fraud. When the frenzy collapsed in April 1864, an important source of capital dried up. To make matters worse, a severe drought and a hard winter hindered wagon freighting on the high plains. Then Cheyenne and Arapaho war parties virtually severed the transportation routes for the best part of a year. Farther east, Confederate guerrillas sometimes attacked wagon trains bound for Colorado. As supplies grew scarcer, the costs of mining in isolated towns grew ever higher. Few producers could operate in such an environment. By the end of 1864 the failure of the stamp mills, the lack of new capital, and a sharp increase in costs had disrupted the mining industry.[11]

As the specter of financial ruin hovered over Colorado, mineowners launched a desperate search for a method of recovering gold from sulfurets. Yet, since scarcely anyone in Colorado possessed a thorough knowledge of mineralogy and metallurgy, the industry fell prey to a group of quasi-scientists who created what Rossiter W. Raymond, the United States commissioner of mining statistics, dubbed the "process mania." The chief objective of the extravagance was to remove sulfur from ores. As Raymond said, "Desulphurization became the abracadabra of the new alchemists." Speaking learned nonsense about "gaseous gold, silver-sheathed gold, air-filmed gold, and chemically-combined gold," these charlatans induced credulous mineowners to purchase contraptions and patent rights that, however absurd, still seemed plausible because they generally employed the familiar technique of amalgamation as the terminal step.[12]

Results were usually disastrous for luckless investors snared by their own desperation, gullibility, and greed. The Mason or Hagan process, for example, allegedly decomposed superheated steam so that elemental hydrogen could attack the sulfurets and free the gold. Dodge's desulphurizer destroyed itself upon use. And the Bartola process was so dubious that Raymond acidly noted that it was "difficult to reconcile the history of this invention with the hypothesis of honesty on the part of the inventor." Yet in spite of their untested nature, interest in the contraptions was intense. Nathaniel S. Keith, later an important figure in the development of electrometallurgy, wrote his wife that curious people came from miles around to see his

"marvelous desulphurizer." But high hopes and large investments went for naught as nearly all inventions were dismal
failures. By 1867 rusting iron machinery visible throughout
the gold region testified to the folly by the "process men."[13]

Early that year a reporter for *Harper's Magazine* wrote that
the "El Dorado of the West" had turned into a "land of disappointment." Throughout the gold region—at Central City,
along South Clear Creek, and in South Park—hard times prevailed. Entire camps lay crumbling into ruin. Mines known to
be rich were boarded and locked. Mills stood "silent as a tomb."
Unused machinery of every kind rusted into uselessness.
Homes lay shuttered and abandoned. And miners' tenements
showed "no smoke, no sound . . no living thing."[14]

Many of those remaining in Colorado despaired of recovery.
The territorial governor spoke with pessimism about the possibility of a revival, and one mineowner later remembered that
everyone who could get away was doing just that. Some predicted the mining towns would soon be deserted. Others
thought the land might even revert to wilderness. What the
mining industry needed, among many things, was a technology
that could recover gold from sulfurets, and it was just about
this time, as the industry collapsed, that a long series of events
were about to produce a solution.[15]

In the spring of 1863, William Gilpin, the mercurial first
governor of the territory, fulfilled a long-held dream by acquiring options to purchase the Sangre de Cristo grant. This was a
million-acre tract of land in the San Luis Valley about 150
miles south of Denver. Gilpin, however, was insolvent as usual
and could not secure the grant outright. To arrange financing,
he traveled to San Francisco, where friends provided him with
letters of introduction to investors in New York. He went east
during the summer and with the help of Morton Fisher, a
former business associate, borrowed $30,000 from the invest-
ment banking house of Duncan, Sherman & Company. To obtain additional funds, he entered into an informal partnership
with Fisher and Colonel William H. Reynolds, a cotton textile
manufacturer in Providence, Rhode Island. With the $41,000
needed for purchase, Gilpin and his partners sought expert

opinion on the mineral resources of their vast domain. Early in 1864 Reynolds approached Nathaniel P. Hill, professor of chemistry at Brown University, in hope that Hill would act as his personal representative in the investigation planned for that summer.[16]

Hill was thirty-two years old at the time and had a fine reputation in both academic and business circles. Born in Montgomery, New York, in 1832, he was the son of a well-to-do farmer who was prominent in state and local politics. He grew up in Montgomery, and after his father's death he managed the farm until 1854, when he entered Brown University with third-year standing. He enrolled in the "select course" in science and a year later was appointed assistant to the professor of chemistry. After his graduation Hill remained at the school teaching chemistry, geology, physiology, and other sciences. His academic advancement was steady, and his contributions to the Chemistry Department were notable. An energetic fund-raiser—an important attribute in a private school—he was largely responsible for obtaining the donations that enabled the university to erect Rogers Hall, one of the most modern science buildings of the time.[17]

In addition to teaching, Hill developed a vigorous consulting business. Mainly, he conducted scientific investigations for local industrialists, but he also analyzed autopsy materials for the city of Providence and once made a controversial study of pollution in the Providence River. Another time he took a leave of absence from the university to operate an oil refinery. Because of his wide-ranging activities, Hill developed numerous business contacts throughout the East, and he acquired a reputation for integrity, thoroughness, and sound judgment that lasted for the rest of his life.[18]

In the winter of 1863 Hill received three offers to investigate mining properties. He regarded the first two as "very advantageous," but he declined them both on the grounds that "the labor required would be incompatible with [his] obligations at the university." The third proposal came from Colonel Reynolds, and it was more lucrative than the others. He promised Hill "very liberal compensation": a salary of more than

Nathaniel P. Hill in the early 1860s, shortly before his first trip to the Rocky Mountains. Brown University Archives.

$2,500, far higher than his university pay, and an option to purchase a tenth of the tract. Hill was to work from the end of the spring semester in April through December. His first impulse was to decline, but, perhaps after Reynolds offered further inducements, he decided to accept if the university would grant him a leave of absence. The executive board granted his request, and two "disinterested" members privately told him not to make an effort to return early if it meant "sacrificing important interests."[19]

There was more to Hill's desire to visit Colorado than what he told the executive board or admitted to Reynolds. Years before, in 1857, he had written a college friend that he considered the West "the most favored section of our country" and that one day he might "emigrate to one of those flourishing cities to seek [his] fortune." He was also intrigued by the investigation itself. He later wrote his wife Alice that he saw it as "one of the finest opportunities to make not only some money, but also fame." And he hoped to invest in mining properties as a means of safeguarding his savings from the steady erosion of depreciating greenbacks. In this he was not alone. When he left Providence, professors Albert Harkness and Alexis Caswell of Brown and Governor Elisha Dyer of Rhode Island gave him money to invest in western mines. Clearly, Hill intended to use his journey for more than the purposes outlined by Reynolds.[20]

At the end of May, Hill boarded an afternoon train at the Providence station, waved good-bye to his family, and set off for the Rocky Mountains. The first leg of his odyssey—to New York, Cleveland, and St. Louis—was by rail, and for most of the way he traveled with a Miss Rathbone, whom he accompanied as far as St. Louis. From there Hill went on alone. He took a steamboat to the port of Hannibal, Missouri, and from there continued by train to Atchison, Kansas, a town he later said was "filled with . . . every kind of abomination." The stagecoach route began at Atchison, and he booked passage for Denver despite rumors of an Indian uprising on the plains. The climate, scenery, and terrain of "the Great American Desert" engaged his attention as the coach bounced along, but he was dismayed by the crudeness of the towns along the way and predicted that "no consideration could induce [him] to live in them."

Gilpin had seen to it that the fourteenth of June would be no ordinary day in Denver City, as the dusty, windy town was still known in 1864. He passed the word of Hill's arrival to the newspapers, notable men, and a few people Hill had known in Providence. By midmorning excitement at the stage depot ran high. Reporters milled about, and one more enterprising than the rest had his editor run the headline A SAVAN AMONG US, a phrase that worried little about observing the niceties of French but befitted a faculty member from Brown University.

Finally, about eleven o'clock a dusty coach came into view, clattered up the street, and screeched to a halt. Out to shake hands with his greeters stepped Nathaniel P. Hill, the man who would do the most to revitalize the foundering mining industry, though no one, not even Hill himself, had any inkling of this.[21]

Despite the fanfare that Gilpin had engineered, the formalities at the station were brief, which was much to Hill's liking, since he was exhausted after his hard, dangerous journey across the plains. He answered a few questions to satisfy the reporters, spoke with an old friend, and exchanged pleasantries with new acquaintances. Then he walked over to his hotel with Gilpin. That night, as he relaxed in his dimly lit room, Hill wrote a letter to his wife Alice. He had arrived in safety despite the Indian uprising on the plains, but seven days and nights of uninterrupted travel in a hard-seated, springless stagecoach had taken their toll. A hot bath and his first change of clothes in two weeks had proved "almost renovating" in themselves. Despite his fatigue, he had lingered earlier to talk with Gilpin about their forthcoming expedition to southern Colorado, an expedition that would explore, survey, and map a tract of land larger than the State of Rhode Island. Hill was to file a report on the mineral resources.[22]

A few days later Hill and Gilpin set out for Central City so that Hill could examine older mining properties and confer with other members of the exploring party. They drove along in Gilpin's buggy pulled by two ponies, "the ugliest, homeliest and meanest rats" Hill had ever seen. Despite these views, which Hill kept to himself, he and Gilpin got along well, but they were very different. Gilpin, a man of medium stature, was now in his late forties (nobody really knew how old he was), his deeply etched face accentuated by a receding hairline that he counterbalanced with a large beard. The long-haired though clean-shaven Hill, a little taller than Gilpin, was nearly a generation younger. Gilpin had established as much of a reputation for bombast as for his achievements as a soldier of the Republic and a mercurial exponent of western expansion. He talked incessantly. Hill seemed introspective by comparison, though he was hardly a quiet person. He found Gilpin amusing and discounted his braggadocio, partly because of his fascination with the environment—the changing weather, the narrow canyons, and the snow-capped mountains that he thought were "grand and sublime beyond description."[23]

In Central City, Hill met other members of the exploring party—Redwood Fisher, James Aborn, and Joe Watson. They insisted that Hill move into bachelors' quarters with them; he accepted and dined luxuriously—so he said—on canned oysters, lobster, and other delicacies. Both Fisher and Aborn were former residents of Providence, and Hill apparently knew them, particularly Aborn, a cousin of Professor John Peirce of Brown University. The day after his arrival Hill visited his first mines, on the Bobtail lode, in company with Henry B. Brastow, who was a business associate of Reynolds and another Providence emigrant Hill had known in former days. He spent the rest of the week looking into mining properties on behalf of Harkness, Caswell, and others, then returned to Denver to join Gilpin's expedition.[24]

The twenty-two-man party got off for the San Luis Valley on the Fourth of July. Gilpin was in overall command, Aborn in charge of the prospectors, Redwood Fisher the surveyor and engineer, and Hill, as he put it, the "Chemist and Mineralogist, Geologist & C." He joked that they were the four "savans" of the expedition. Gilpin had seen to it that his men were well armed and equipped. They had two wagons and an ambulance to carry medicines, tools, and scientific equipment, as well as an arsenal of pistols, rifles, and shotguns for hunting game and repelling any hostile Indians who might challenge their passage. Nonetheless, Hill belittled Gilpin's concern over hostilities raging on the plains, and perhaps Hill was right. Except for Aborn's recurrent nightmares about snakes, the journey from Denver was uneventful. They reached Fort Garland, northern outpost of the grant, a week later.[25]

Once the investigation began, Hill decided to modify his plans when he found he would be idle "nine-tenths" of the time. He had originally intended to spend four weeks on the tract and then head north to inspect mining properties near Boulder

Central City, Colorado, in 1864. Note the denuded hillsides. George D. Wakely, photographer. Colorado Historical Society, Denver.

and Central City. After that he would rejoin the expedition to examine what ore deposits Gilpin's men had located in the interim. But Hill was unwilling to do nothing while Aborn's prospectors were searching for minerals, and so he seized an opportunity to return to Denver with two hard-drinking army men traveling from Fort Garland.[26]

The next few weeks were fruitful—or at least potentially so. Hill examined new ore discoveries in the hills above Boulder, then went to Central City, where he purchased options on what he thought were valuable mines on the Bobtail, Fairfield, and other lodes. He also used a portion of his salary from Reynolds to buy a house and lot in case he should ever move to the high country. On his way to rejoin Gilpin's party in September, he wrote his wife that he was very pleased with his trip so far, although its success depended upon the "pecuniary results" of business transactions yet to come. They looked promising, but they still might fail.[27]

Hill spent only a few weeks on the Sangre de Cristo grant that fall. With the assay equipment he had carried from Providence, he tested ore samples brought in by Aborn's prospectors, recorded his findings, then returned to the "little kingdom of

Gilpin" to continue his investigation of the mining industry that by now consumed his interest.

Once in Central City, he learned that several friends had departed the previous day to inspect new ore discoveries at Red Mountain, a peak west of the Arkansas River, about ninety miles away. Hill was anxious to catch up because some of the new mines were said to be the richest in the world. The next morning he set out on horseback with the Reverend T. D. Marsh, a Methodist minister who accompanied him as far as Fairplay. From there Hill traveled alone, crossing the mountains from South Park to the Arkansas Valley in the midst of snow flurries, a sign of early winter in the high country. He caught up with his friends at Twin Lakes, a few miles west of the river, and went on with them to inspect the lodes at Red Mountain. On their return to Fairplay, they were caught on the high, bare ridges by a severe snowstorm and only with difficulty managed to reach the valley floor. When the weather cleared, Hill and a Mr. Byram rode on to Montgomery and Buckskin Joe, where they examined mining properties for two days before returning to Central City in the midst of another snowstorm.[28]

By now it was time to depart. With an early winter setting in, Hill wound up his business affairs and made the usual round of good-byes. A local newspaper reported that he was "agreeably disappointed with the character of the country" and considered it "eminently worthy of another visit." At the end of October he left Denver by stagecoach, this time in company with William A. Abbe of Boston, a new associate whom he described as "a clever fellow with plenty of means." They arrived in New England one month later.[29]

Once back in Providence, Hill submitted his report to Colonel Reynolds. This document never became public—and no copies have come to light—but it was widely rumored that Hill had found little of value in the San Luis Valley. Circumstantial evidence tends to corroborate this view. Reynolds sold his interest a few months later, and Hill never exercised his option to acquire a one-tenth interest. Five years later, in 1869, Gilpin and his partners published a promotional brochure containing a letter from Aborn claiming that the grant was as rich in minerals as Central City. If Hill had made a similar report, it would surely have appeared with Aborn's letter.[30]

Having discharged his obligations to Reynolds, Hill confronted the question of his future. Should he remain as professor of chemistry at Brown University, or should he become an industrialist? This was not an easy decision. On his travels in Colorado he had missed university life, and he had looked forward to rejoining his friends and colleagues. Yet about the time he left Central City he had grown apprehensive about his position. Somehow he had gotten wind of trouble in Providence. He wrote his wife that the "college fraternity" may have ruled him out of its ranks because "certain gentlemen of the faculty" did not appreciate his neglect of the university for the "base motive of getting money." He had hoped to smooth over any ruffled feelings and resume his academic career, but once in Providence Hill realized that he was now so committed to business that he could no longer lead the double life of professor and entrepreneur. In November he submitted his resignation to the executive board.[31]

As the drizzle, rain, and snow of a New England winter closed down on Providence, Hill disposed of the options he had taken on mines at Central City. He had little trouble in doing this, but he retained two options. With these he organized the Sterling Gold Mining Company and the Hill Gold Mining Company. Then, eager to begin operations, he left for Central City in February 1865.[32]

On the long trip west he traveled with John H. Barlow, a Bostonian he had hired as the agent, or manager, of the Hill Company. At Atchison they purchased tickets on a stagecoach "for such a ride," Hill wrote, "as but few people would be willing to undertake." They knew that Indian war parties were ranging over the route and that drifting snows had obscured the trail. They had no idea how far the driver could take them through subzero temperatures, whether they could fall in with a wagon train, or whether they would meet a military escort. En route they passed fresh graves and burned-out way stations and heard tales of narrow escapes from arrow and lance. Even the military proved dangerous. At one stop they had to guard the stagecoach from drunken soldiers who threatened to steal their belongings. But Hill's luck still held. He and Barlow reached Denver City in safety and then went up to "Central." Here they were joined by E. C. Gould, who had suffered through an equally hazardous passage on his way west to become agent for the Sterling Gold Mining Company.[33]

Once he rented an office, Hill set to work with Gould and Barlow to begin production at the mines and build a stamp mill on the Sterling property. He had complete confidence in Barlow, but he had reservations about Gould. Hill was nonetheless elated. He wrote his wife that he was confident of "splendid results" and was "willing to live or die" with both investments. As operations rolled along in March, he thought the Sterling Company might pay a 25 or even a 35 percent dividend the first year, and a few days later he predicted that both firms would rise 100 percent above their purchase price. He also took satisfaction in the mines he had obtained for eastern investors, and he prided himself on his new reputation as a superior judge of mining property.[34]

Optimism notwithstanding, Hill's enterprises soon foundered on reality. As spring gave way to summer, Gould and Barlow exhausted the layer of weathered upper ores that were easily processed by stamp milling. Below lay a great tonnage of sulfurets, the copper and iron pyrites creating havoc throughout the region. When Hill sent them to his mill, the yield of gold fell as precipitously as the stamps. Instead of enjoying a large dividend, he now faced the unpleasant prospect of personal embarrassment and financial disaster. Yet he was not deluded by the "process men" because he was himself a real chemist, one of few in the mining country. Instead, he pondered the seemingly insoluble riddle that was carrying a thriving industry into depression. It was about this time that he became involved in a project that hoped to recover gold by a method known as smelting—a complex high-heat process in which copper, lead, or other base metals in an ore served as a vehicle to collect and hold the precious metals.[35]

The initiator of the scheme was James E. Lyon, a convivial

though somewhat irascible entrepreneur whose long involvement with mines and mills at Central City had thoroughly acquainted him with sulfurets. He may have considered the possibility of treating them by the smelting process as early as 1862, well before the problem became acute, but he did nothing for three years until 1865, when he shipped some ores from the Gregory lode to a smelter on Staten Island in New York City. There he watched the furnaces recover about $250 in gold from every ton—far more than stamp mills could obtain with mercury. Lyon was elated. Convinced the process would work in Colorado, he persuaded three eastern capitalists, Charles H. Moore, Elisha Mack, and Fulton Cutting, to build a smelter as an adjunct to the New York & Colorado Mining Company, an enterprise in which they had invested with Lyon and his western partner, George M. Pullman, later of palace car fame. Using some of the money, Lyon purchased equipment on Staten Island and hired Franklin Johnson and his brother Charles, two knowledgeable European metallurgists who had emigrated to the United States. Lyon also paid the passage to Colorado of several experienced smelter workers after they had agreed to remain in his employ for two years.[36]

Once he returned to Central City, Lyon engaged Hill as a consultant on the project. Why he did this, particularly with experienced metallurgists on the way, is certainly curious, because Hill had little knowledge of smelting processes. The two men knew each other at least casually, for they had met the year before, and even now Hill was running his mines from an office in one of Lyon's buildings. Perhaps Lyon needed a skillful chemist, or perhaps he valued the advice of a well-educated man. Whatever the reason, Hill acquired his first practical introduction to the smelting business that summer.

Lyon planned to install the machinery essential in smelting silver-lead ores and to integrate this later with the furnaces needed to reduce sulfurets. In August he erected a small experimental plant in Black Hawk and with the Johnson brothers set up a Scotch hearth and a cupel furnace to produce and refine silver-lead bullion. When the Johnsons turned out a silver ingot, reportedly the first ever cast in Colorado, the

ORE DRESSING ROOM – THE BUDDLE AND JIGS.

REVERBERATORY FURNACE.

CUPEL FURNACE.

SCOTCH HEARTHS.

Inside the smelter built by James E. Lyon. A. E. Mathews, delineator. J. Bien, lithographer. Colorado Historical Society, Denver.

Daily Miner's Register of Black Hawk hailed the "complete success" of the experiment and proclaimed the arrival of the "millennium" for the depressed mining industry—somewhat prematurely, as events would prove.[37]

Lyon was so encouraged that he hurried the project along. He had the Johnson brothers install crushers, assaying equipment, and reverberatory furnaces, and he purchased pyrites and tailings to supplement the production of his own mines. In December the plant turned out its first bar of gold, prompting the *Rocky Mountain News* to proclaim "Victory! The Day Is Ours! The Lyon Process Successful! 500 Pounds of Bullion in the First Run!" A week later, the *News* had to modify its exaggerated estimate of the output when a correspondent discovered that the bullion weighed only one-tenth as much as first reported. When a series of technological difficulties proved the *News* had been premature in hailing Lyon's success, the editors sheepishly backtracked, claiming they had been careful not to commit themselves unguardedly.[38]

While Lyon pushed forward with his project in the fall of 1865, Hill returned home to Providence to launch his own study of ore reduction. He was by no means an expert in metallurgy, but he had some knowledge of the techniques employed, which served as his starting point. He drew up a list of several possible technologies, then whittled down the number by discarding one method after another. By November he was convinced that he should adopt a Welsh smelting technique known as the Swansea process. In this method the copper and iron sulfides in the ores could be used as a vehicle to collect and hold the gold and silver in a mixture known as a copper matte. This product could then be shipped to wherever it could be separated (or refined, in metallurgical terms) into gold, silver, and copper metal.[39]

It was only a short step from identifying a possible technology to the idea of building his own smelter, but Hill knew such an enterprise would be risky. There were no successful smelters anywhere in the intermontane West, and the Swansea process was almost wholly untried in North America. He realized that his scheme called for careful planning to attract venture capital and provide the optimal chance for success. Rather than plunge forward as Lyon was doing, he decided to go to Europe to observe the Swansea process in operation and discuss the entire metallurgical question with experts.

During the winter he obtained letters of introduction, and in February 1866 he set sail for Great Britain on board the *Europa*. This time he traveled with Charles W. Lippitt, one of his former students and a future governor of Rhode Island. They suffered together through a stormy Atlantic crossing but reached Liverpool in safety and went on to London by train. There Hill talked with Sir Edward Sabine, president of the Royal Society, and with Dr. John Percy, a world-renowned authority on nonferrous metallurgy. These conversations convinced Hill that the Swansea process was the correct method for treating the sulfurets of Gilpin County.[40]

From London, Hill and Lippitt went on to the world-famous smelting center at Swansea, Wales. Situated between the coast and the coalfields, this community was the chief center of the British copper industry. The smelters drew ores from around the globe, and because the two held such eminence, it attracted a long line of entrepreneurs in search of technology. Nathaniel P. Hill was neither the first nor the last. Despite its significance, Swansea was a bleak place to live and work. When Hill arrived in 1866 he found "a filthy, crowded, smoky dingy town." Sulfurous fumes had choked every bit of vegetation, and smoke from the smelters hung like a pall over the whole area.

Though Swansea resembled the drab scene of a Dickens novel, Hill spent three valuable days touring plants, visiting coal mines, and talking with metallurgists. He spent much of his last day with B. J. Herrmann, a chemist at Vivian & Sons, the largest company in town. They discussed Hill's plans to erect a smelter in Colorado and arranged for a joint trip to the high country. After leaving Wales, Hill made a quick journey across the English Channel to examine ore dressing machinery at Aix-la-Chapelle (now Aachen), then returned with Lippitt to the United States.[41]

While Hill was tapping the best scientific knowledge in Europe, Lyon continued his efforts to establish a successful smelter in Colorado. Early in 1866 he went to New York for a meeting of the stockholders of the New York & Colorado enterprise. The works at Black Hawk were not in satisfactory operation despite expenditures of nearly $225,000, more than two-thirds of it for construction. Yet Lyon still remained hopeful about the project. Largely through his efforts the firm was reorganized as the Pioneer Smelting Company, and new capital was subscribed. In the spring he discussed the mechanics of the Swansea process with Herrmann, who had arrived in the United States and was en route to Colorado with Hill.

As a result of his conversations with Herrmann, Lyon abandoned the lead-base technology in favor of the Swansea process. When he resumed operations, he obtained about a hundred tons of copper matte that he shipped to Wales for refining. But in spite of this measure of technical success, profits remained elusive and the enterprise failed. Early in 1867 the plant was sold. When workers dismantled the furnaces, they found another hundred tons of matte that had leaked undetected through the hearth into the masonry. Rossiter Raymond claimed Lyon could have made a profit if the loss had been discovered and the product shipped.[42]

Lyon failed to establish a successful smelter because he employed many different methods instead of concentrating on a single, standard process. He wasted time and money on technology that was not applicable to most of the ores mined in the vicinity of Central City. In spite of the alleged expertise of the Johnson brothers, his plant was inefficiently designed, badly built, and ill-managed. His efforts nevertheless showed that the smelting method could recover gold and silver from sulfurets, and he was the first to employ the Swansea process in Colorado. His successors, Nathaniel P. Hill in particular, profited by his failure, but this must have been little solace to Lyon.

In the meantime Hill continued his thorough investigation of the Swansea process. Instead of plunging ahead and building a smelter, as Lyon had done, he was more careful, knowing the risks involved. He wanted to ship a large quantity of ore to Wales, then go there to observe the actual treatment. This, of course, required a large capital outlay, one that might not prove remunerative to investors. Despite the risk, Hill obtained support from J. Warren Merrill, a Massachusetts chemical manufacturer who had invested in Colorado's mines the year before. Others joined in the project, thus spreading the risk.

With the plans set, Hill met Herrmann in New York. The two men traveled to Colorado, where the Welshman examined mines and assayed samples of ore. He called stamp milling "a most wasteful expenditure" and termed smelting "the only practicable solution" to the problem of recovering gold from sulfurets. On his advice Hill spent $7,000 to purchase seventy tons of mineral from the Trust mine on the Bobtail lode. During the summer and fall he had the ore freighted across the plains to Atchison, floated down the Missouri and Mississippi rivers to New Orleans, and shipped to Vivian & Sons in Swansea. In the fall Hill made his second voyage abroad, this time to observe treatment. The process worked well, and he began formulating plans to erect works in Colorado. In spite of its success the experiment cost Hill and his associates $19,000 leaving them with a $400 loss.[43]

Early in 1867 Hill obtained financial support from easterners interested in the potential of a reduction enterprise. Together they organized the Boston and Colorado Smelting Company. On May 11 the articles of incorporation were filed in Boston by James W. Converse, J. Warren Merrill, Joseph Sawyer, and Gardner Colby, who set the capitalization at $200,000. Hill was named agent and local manager of the corporation.[44]

The founders were prominent businessmen in the Boston area. James W. Converse, president of the firm for many years to come, came from a family of shoe manufacturers and was a speculator in western enterprises. J. Warren Merrill held a seat in the general court of Massachusetts and once served as mayor of Cambridge in addition to being a chemical manufacturer. Gardner Colby was an importer of dry goods, a manufacturer of woolens, and, like Converse, a speculator in western ventures. Joseph Sawyer was another woolen producer and mineowner. Hill apparently met Colby and Merrill through Brown University, for both were trustees of that school as well as of what are now Colby College and Andover-Newton Theological Seminary. Several of Colby's sons had also studied under Hill. Perhaps he met Converse and Sawyer through them or through his other associates in Boston. Regardless, Hill and his colleagues had one interest in common: a desire to invest in western enterprises, whether in mining, ranching, or railroad building. The Boston and Colorado Smelting Company was the type of venture that appealed to them, but even as they set their hands and seals on the legal documents, two federal mining engineers were reporting that there was "reason to believe that the proper economic conditions for smelting do not exist" in the Rocky Mountains.[45]

Chapter 2

High Country Years

EVER SINCE HIS FIRST ARRIVAL IN 1864, COLORADANS HAD FOL-
lowed Hill's activities with interest. He was respected as both a
scientist and an entrepreneur, and people watched his efforts
to find a solution to the riddle of the sulfurets. But when the
news arrived that he was about to build a smelter, many were
skeptical. James E. Lyon had just failed, and here was Professor Hill about to try the same method. The "process mania" had
returned again, many thought.

Hill and his associates had no illusions about the speculative
nature of the Boston and Colorado Smelting Company, and
they laid careful plans. At meetings in New England, they
decided to construct a small plant that would have two roasting
furnaces and one smelting unit to produce copper matte.
Though it would be small by Welsh standards, the furnaces
would be the largest size in use, so that Hill could take advantage of the economies of scale, an important consideration in a
high-cost mining region. But Hill and his colleagues decided
not to erect a refinery, since the smelter would not produce
enough matte to make separation economically feasible. Instead, they negotiated a contract with Vivian & Sons and pre-

pared to ship the matte to Swansea. Clearly, Hill and his associates wished to conserve their limited capital resources and minimize whatever losses they would have to absorb if so speculative an enterprise should fail.[1]

The founders of the company also wanted an experienced metallurgist to handle all technical matters, for everyone recognized that Hill was not expert in the details of the Swansea process despite his two trips abroad. The firm thus turned again to Europe, where it hired Hermann Beeger, a crusty German metallurgist whom Mrs. Hill always referred to as "that old Dutchman." He had learned his trade at the Bergakademie of Freiberg, Saxony, Europe's most prestigious mining school, and had worked at smelters on the Continent and in Swansea. Hill also hired a few skilled workers in Wales and paid their passage to Colorado.[2]

Finally, Hill and his associates had to select a location for the plant. They realized that the ideal site would be a railroad junction at the base of the mountains. Over an efficient system of interconnecting lines, ore cars from different mining districts could roll downhill to the reduction furnaces, and coal could be brought in from southern Colorado. But this was simply not possible in 1867. The territory was still an undeveloped region. It had no rail service of any kind, and what wagon freighting existed was chronically inefficient, vulnerable to the vagaries of drifting snow and quagmires of mud. The southern coalfields were virtually untouched. With their options circumscribed by the lack of industrial development, Hill and his colleagues had to erect the smelter in the midst of its supplies of fuel and ore. This meant Gilpin County, the heart of the mining region. Once Hill had moved his family to Colorado, he purchased a four-acre parcel of land in Black Hawk near the site of Lyon's defunct smelter.

Dirt flew along the banks of North Clear Creek that summer. Hill hired construction crews to dig out foundations and put up buildings. He purchased engines, boilers, and other equipment secondhand from local mining companies, but iron castings had to be shipped across the plains from St. Louis and firebrick hauled up Clear Creek Canyon from Golden. Beeger directed the installation of the machinery, supervised furnace construction, and took charge of all other metallurgical details.

Once the furnaces had been completed, Hill and Beeger launched a series of experiments, only to encounter a serious, wholly unforeseen problem. When the ore hopper suspended over the smelting unit dropped an ore charge into the hot furnace below, the firebrick lining the walls cracked—unable to withstand the sudden, extreme temperature change. This forced Beeger to redesign the unit so that ores could be added through a charging door he had installed in the side. The alteration decreased efficiency and added to the costs of operation, but there was no other way to resolve the difficulty. High-quality firebrick was unavailable at any price. To their chagrin, Hill and Beeger had to abandon the new Gerstenhofer roaster for the same reason; but, much to their relief, the roasting furnace built along traditional Welsh lines functioned properly.[3]

Everything was now ready. With the experiments successful, Hill and Beeger designed an ore-purchasing formula based upon policies in use at Swansea, since there were no adequate precedents in Colorado. With this, Hill signed contracts with the Gregory, Bobtail, and other mining companies. The ore bins filled up, and in February 1868 the Boston and Colorado Smelting Company began its commercial operations.

There were two steps in the Swansea process as adapted by Hill and Beeger. At first, they roasted both ores and tailings to reduce the sulfur content, then smelted the intermediate products to obtain copper matte. The technology was complex, but it was relatively unsophisticated compared to what it became in future years.

Hill and Beeger used two roasting methods. In one, workmen piled ores upon specially constructed wooden platforms that were ignited to bring the pyrites to their own point of combustion. Once on fire, they were allowed to burn for six weeks in the open air, generating clouds of pungent, sulfur-dioxide-containing smoke that darkened the skies above Black Hawk. Mill tailings, however, went through a different roasting. The stamps had pulverized them so finely that they would not burn by themselves; Beeger had to roast them for twenty-four hours in a furnace known as a calciner. In either case the preliminary treatment reduced the sulfur content of the minerals to about 4 percent, making them ready for smelting.[4]

The smelting unit was a long, flat structure known as a reverberatory furnace. The working area was an oval hearth covered with a low, arching roof. At one end lay a sunken firebox separated from the hearth by a projection known as a firebridge. This design was the key to the process. The curvature of the ceiling bent the flame at a right angle to its natural course, distributing heat and burning gases evenly over the hearth, where temperatures reached 1,400 degrees centigrade. The flame that followed the underside of the arch for nearly the length of the furnace was said to reverberate its heat downward onto the hearth. This gave the unit its name.[5]

Smelting was continuous. At the start Beeger prepared a mixture of ores and tailings weighing about two tons, which was pushed onto the hearth through the side or charging doors. During the next six to eight hours the furnace melted this material, while smelter workers periodically agitated the molten mass with long iron rakes known as rabbles. Because of their high specific gravities and their affinity for one another, gold, silver, and copper, together with iron and copper sulfides, settled to the bottom of the hearth as a dark, lustrous liquid that was easily distinguished from the worthless slag that coalesced above. Sulfur dioxide gas escaped through a chimney, to the detriment of Black Hawk's atmosphere. As the separation neared completion, Beeger assayed the molten slag. Once he found it free of precious metals, it was skimmed from the surface of the matte, allowed to cool, and discarded on a dump. After several charges had been treated in this fashion, smelter workers "tapped" the matte through an aperture opposite the charging door. Once the matte solidified, it was crushed, assayed, sacked, and shipped to Swansea for refining. James D. Hague, a shrewd mining engineer who visited the plant soon after operations began, reported that the loss of precious metals in the entire process was less than 5 percent of the charge, an excellent figure for that time.[6]

Despite his careful planning, Hill found the Swansea process expensive. When he went into business, the ridges surrounding Black Hawk had already been stripped of trees to satisfy the voracious demands of mining companies. To fuel his furnaces, he had to import wood at prices ranging up to six dollars a cord. Labor costs were equally high. Hill paid his skilled workers from five to six dollars daily, the unskilled about half that amount. Everyone worked a twelve-hour shift, six days a week. Fuel and labor costs pushed the price of reducing a ton of ore to about thirty-five dollars. On top of this, Hill had to ship the matte to Swansea for refining, which added another fifteen dollars to the cost of processing. And, finally, there was the capital tied up in ores and matte, an estimated $200,000 by the end of 1870.[7]

As might have been expected, Hill's ore-purchasing schedule reflected his smelting costs. He paid for gold on a sliding scale. For ores holding two ounces per ton, he paid only one-fifth of the metal's value—twenty dollars a ton in 1868. This return—only four dollars an ounce—seemed niggardly to many, but the cost of reduction was more than thirty-five dollars a ton, and a two-ounce ore without copper and silver could be smelted only at a loss. As the quantity of gold rose, however, so did the purchase price, until it reached a ceiling at 60 percent. Hill purchased copper and silver on a different basis. He paid for copper on a straight percentage scale minus a fee to compensate for losses in assaying. The silver schedule was more involved. For each ounce, worth about $1.29 on the open market, he returned seventy-five cents, but only after the percentage of copper had been subtracted from the number of ounces of silver. In other words, if the numerical percentage of copper exceeded the quantity of silver, mineowners received nothing for the silver. Such instances were rare, and Hill was flexible in dealing with mining companies. Shippers with exceptionally rich ores often bargained for higher returns by threatening to send their production to Swansea or to the rival smelters soon built in the Rockies.[8]

Hill drew his ores almost entirely from local companies. Mines on the Gregory, Bobtail, and Gunnell lodes produced

Fig. 1. Cross-sectional view of a calcining furnace. Tailings were placed on the uppermost hearth (near B) and gradually advanced to the lowest and hottest tier adjacent to the firebox (near A). Reproduced from Egleston, "Boston and Colorado Smelting Works."

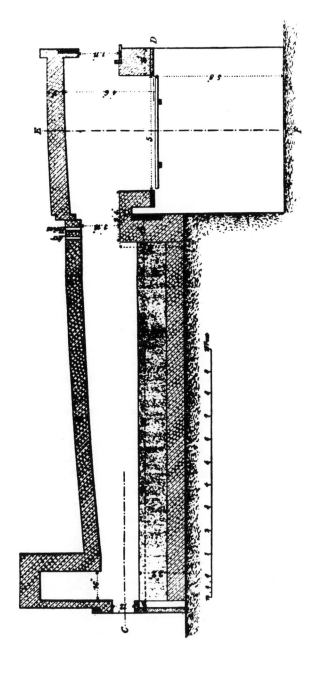

Fig. 2. Cross-sectional view of a reverberatory furnace. Note the sunken firebox on the right, the firebridge below the airholes, the long hearth and curved ceiling, and the chimney at the extreme left. Reproduced from Egleston, "Boston and Colorado Smelting Works."

large quantities of sulfurets averaging five ounces of gold, ten ounces of silver, and 5 percent copper per ton. Some shipments were very rich. When prospectors located silver ores in Slaughterhouse Gulch near Black Hawk, Hill purchased some holding as much as 270 ounces per ton. The value of mill tailings varied. The pulverized rock sent from the Burroughs and Bobtail lodes was worth almost as much as the unprocessed ore, continued testimony to the failure of stamp milling.

Hill got the company off to an auspicious start. Despite the bad firebrick, the failure of one roaster, and high costs, the Swansea process worked well on sulfurets. Clouds of smoke drifted skyward from the great roasting piles, molten ores hissed in the furnaces, and the clang and crash of ores and heavy machinery raised a din around the plant. Yet it became obvious in a very short time that the method worked. Of this there was visible evidence. Beeger reduced the ores into a matte averaging fifty ounces of gold, one hundred to two hundred ounces of silver, and 50 percent copper per ton, each one worth about $2,000. Hill shipped the first lot to Vivian & Sons in June, four months after operations began.[9]

Despite the propitious start, the smelter nearly came to a premature demise on June 17. About four o'clock in the afternoon, while workmen were skimming slag from the smelting furnace, a burst of sparks escaped from the firebox and ignited the roof. In minutes the building was in flames. Alarms rang out through Black Hawk and brought a motley collection of clerks, storekeepers, miners, and other volunteers to the scene, but before they could organize bucket brigades, a light wind blowing through the canyon carried the blaze to other structures. By the time the firefighters had extinguished the last ember, much of the plant lay in ruins. That night Hill sadly estimated the damage at $8,000 to $10,000. More pessimistic observers feared the works a total loss.

The next morning Hill came down to examine the plant again. He poked through the debris with Beeger and found that the furnaces had survived the fire intact. With that, he revised his judgment of the loss to the far more modest figure of $3,000. He announced that the firm would rebuild, extended his "heartfelt thanks" to the people of Black Hawk for their "magnanimity" in putting out the blaze, and let contracts to rebuild the shattered plant. A few weeks later the smelter resumed operations.[10]

The fire proved only a temporary setback, and in the months that followed, "Professor Hill's works" provided a local ore market that stimulated a revival of mining throughout Gilpin County. As new capital flowed in, mineowners retimbered old shafts, installed new pumping machinery, and gradually reopened properties that had sat idle for years. A new spirit of enterprise prevailed. Old smelters in Wales and new smelters in Colorado competed for the rising production, but Hill proved himself skillful in business. By 1871 Anton Eilers, the friend and deputy of Rossiter Raymond, noted that Hill had virtually monopolized the smelting ores mined in the "little kingdom of Gilpin." Years later Hill himself reminisced about this time when the mines produced only two kinds of mineral—the "Hill ore" and the "mill ore."[11]

But Hill and his colleagues never engaged in nostalgia in those early years—they increased their smelting capacity to handle the rising output. The first expansion came in 1869 when Beeger erected a second calciner and a second reverberatory furnace, both duplicates of the original models. The next year he put two more calcining units into operation, and in 1871 a third reverberatory. They increased the ore consumption of the works to twenty-five tons daily, triple the capacity of the original plant. Hill and his colleagues also purchased a rival smelter erected in Black Hawk by William West, a British-trained mining engineer.[12]

Secure in the ore markets of Gilpin County, Hill and his colleagues looked abroad. Over the mountains to the south lay Clear Creek County, where prospectors had discovered silver on the crags that towered above the hamlets of Empire, Idaho Springs, and Georgetown. Only small quantities of mineral had been shipped, owing to the high transportation costs and the lack of an ore market. The bulk of the mineral was galena unsuited to the Swansea process, but some mines held large quantities of pyrite and low-lead rock. This was what Hill wanted. He made his first effort in the county in 1872 when he bought two thousand tons of ore and had them hauled over the mountains to Black Hawk despite a huge freight charge of fifteen dollars a ton.[13]

Unlike the easy path he followed to domination in Gilpin County, Hill met real competition when he entered the markets of South Clear Creek. Welsh firms like the Vivians purchased mineral in the valley even though the shipping costs exceeded $125 from mine to furnace. British investors also organized several companies that erected local plants using the Swansea process. Yet the mines of South Clear Creek did not produce enough ore to supply so many smelters. Hill competed effectively, helped by his well-established monopoly in Gilpin County. The British companies proved unprofitable, and to seal their doom Hill hired their metallurgists, Richard Pearce and Henry Williams. The smelters in Wales remained marginally competitive for a time, but when railroad construction finally lowered Hill's costs, he increased his returns above the pay of the Swansea works and drove them from the ore market.[14]

Hill also tapped mines to the north in Boulder County, particularly near the boom town of Caribou, whose silver leaped to fame in the early seventies. Mills at Nederland and other spots garnered some ores, but mineowners sent large shipments of mineral over the tortuous mountain roads to the works at Black Hawk. During the bonanza days, William West considered erecting a smelter at Caribou, but he wisely refrained, for the local production scarcely justified it and he would have been hard pressed to compete with Hill's company. Caribou's glory flickered later in the decade, and Boulder County never came to rival its southern neighbor as a major mining region.[15]

Despite their success in helping revitalize the mining industry, Hill and his colleagues still found themselves the target of complaint. In 1868, only two months after they commenced operations, the *Rocky Mountain News* of Denver reported grumblings that smelting charges were too high, treatment too slow, and the work unsatisfactory. But these reports were only the beginning. The newpapers later told of other grievances regarding the complicated price schedule, Hill's alleged cupidity, and the supposedly astronomical profits of the enterprise. In 1869, for example, Robert O. Old, a promoter trying to entice British capital into the smelting industry, published a circular claiming that Hill's firm was earning $50,000 a month, or $600,000 a year. And then John H. Tice, a journalist from Missouri who traveled through the high country in 1871, wrote a book that crowed that returns from the smelter had already made Hill a millionaire. Such claims, however, were absurd. Hill's annual production did not exceed $600,000 until 1870, and total output for the first three years came to only $1.4 million. Hill took issue with his critics. In letters to local newspaper editors and to the prestigious London *Mining Journal*, he charged that Old and others had made statements "calculated to mislead" investors, and he challenged potential rivals to reduce ores at lower cost than the Black Hawk works.[16]

Hill received some support in the controversy. In one of his annual reports on western mining, Rossiter Raymond excoriated the practices of some unscrupulous smelter operators, but he thought that Hill's policies were aboveboard and his charges reasonable. Raymond took care to point out that Hill lost money in smelting low-grade ores and had to make up the difference on richer minerals. Even the *Rocky Mountain News*, which nearly always criticized Hill, conceded that smelting was a risky business, the company's profits smaller than generally thought, and the technological and financial problems largely misunderstood. And, the editors noted, the Boston and Colorado enterprise was performing a great service to the territory and its principal industry.[17]

Such arguments, however, did little to assuage critical mineowners. They wanted a greater return on their ores, a simpler pricing schedule, and a competitive smelting industry. This is what they received in the years ahead, but even then there was little harmony. In the struggle for profit, the relations between miners and smelters usually ranged from lukewarm to strident. Neither Hill nor his future rivals ever stemmed the complaints from the mining community. It was a question of economic self-interest, and smeltermen who engaged in mining could be on both sides of the question at once.

As the company expanded, Hill found the volume of work too

great to handle by himself. He needed an assistant, and in June 1870 he hired Henry R. Wolcott, a young man with delicate features, "to keep the books, attend to the office, and do some of the outside business." The older brother of Colorado's future senator Edward O. Wolcott, Henry came from an old but prominent New England family, and he had studied at what is now the Massachusetts Institute of Technology. He had arrived in the high country the year before to try his hand at mining and milling, but he had had little success and was glad to join the firm. He quickly emerged as one of Hill's most trusted associates and became assistant manager of the Black Hawk plant. None of this came too soon, for scarcely a year after Wolcott joined the company, a resurgence of mining in South Park offered new opportunities for Hill and his associates.[18]

The mining industry in South Park had followed the boom and bust pattern characteristic of the high country. Fifty-niners had located placer deposits and lode mines that brought hundreds of treasure seekers to the slopes of the Mosquito Range. Hard-rock mining prospered until the free-milling ores gave way to sulfurets. This brought on a depression, for local mineowners were no more able to recover gold than their counterparts along the eastern slope. As jobs gave out, people drifted away. In 1870 Raymond reported only one mining company operating in the county. Yet the fortunes of South Park shifted abruptly the next year when prospectors located silver-bearing ores on the flanks of Mount Lincoln and Mount Bross. Many discoveries lay on the sites of abandoned gold claims, and litigation over titles offered a bonanza for lawyers. Yet the discovery of silver provided a powerful impetus to the mining industry. Mills were built to handle low-grade ores, but richer minerals had to be smelted to recover the silver. That first summer one mining firm shipped about thirty tons to Swansea.[19]

South Park silver prompted the immediate attention of the Black Hawk smeltermen. In September 1871 Hill, Beeger, Wolcott, and two others purchased a group of placer and lode claims near the base of Mount Lincoln and Mount Bross. Joseph A. Thatcher, a Central City banker, held the properties in trust for eight months while the partners organized the Alma Pool Association to develop the mines. Hill knew that the Black Hawk works could process the ores, but he also knew that hauling them over the mountains was prohibitively expensive. Since the prospects for South Park appeared excellent, he concluded that the future production of pyrites could supply a local smelter. During the winter he persuaded the officers of the Boston and Colorado firm to build a branch at Alma, a small town between Mount Lincoln and Mount Bross.[20]

With new money forthcoming through an increase in capitalization, Beeger and Wolcott moved to Alma, where they erected a plant similar to the original works in Black Hawk. Wolcott took charge of business matters, coordinated operations with the mines of the Pool Association, and signed contracts with other firms. Beeger smelted the ores into a rich silver matte that was shipped to Black Hawk for assaying and to Swansea for refining. With the pioneer smelter in South Park, Wolcott assured the firm of large ore supplies before rivals entered the market—and they were not long in coming. In 1872 Edward D. Peters, Jr., erected the Mount Lincoln Smelting Works at the town of Dudley. At first he used a lead-based process for treating galena, but later he switched to the Swansea method for reducing pyrites. The change came too late, however, for by then Wolcott had nearly monopolized local production for the Alma smelter. Peters's enterprise failed. By the mid-seventies Wolcott and Beeger were treating as much as 90 percent of South Park's output.[21]

Even though they had made a success of the new smelter, Hill and his associates still had some annoying problems. The high costs of fuel and freight cut into profit margins just as they did at Black Hawk; but particularly vexatious was a feud that developed between Hill, Beeger, and Wolcott, on the one hand, and their partners in the mining company, by now reorganized as the Park Pool Association. The source of the friction was the high cost of reduction. William H. Stevens, a large stockholder, charged Hill, Beeger, and Wolcott with using the monopolistic position of the smelter to garner the lion's share of the profits from the mines. Angry letters flowed back and forth. Wolcott denied the accusations, but he failed to soothe Stevens's feelings. The two sides never resolved their differences, and the dispute continued for several years until both parties sold out.[22]

Despite the verbal altercation, the Alma plant continued to run smoothly, though its management changed. In 1874 Beeger left to take charge of the metallurgy at a refinery the firm built in Boston as part of its continuing efforts to integrate operations. After he left, Wolcott assumed the technical and managerial responsibilities for a year until he returned to Black Hawk. Hill replaced him with Henry Williams, a slender Cornish engineer who had managed one of the unsuccessful British smelters on South Clear Creek. During this time the Alma plant served its purpose well by permitting Hill to tap a lucrative ore market he could not have entered in the era of bad transportation. And by providing a smelter for the ores mined by the Pool Association, the plant offered Hill, Beeger, and Wolcott a degree of informal integration that was characteristic of entrepreneurs in the smelting industry.[23]

As Hill and his associates expanded their smelting operations, another significant development took place in Colorado—railroad construction. In 1870 the Kansas Pacific line ran its first train into Denver: "the finest day in the history of Denver," according to one booster. The same year the Colorado Central and Denver Pacific roads connected Denver to Cheyenne, which sat on the main line of the Union Pacific Railroad. Once these firms linked Colorado to regions east and west, the railway system in the territory grew rapidly. The Denver and Rio Grande Railway built south, connecting Denver to the coalfields at Canon City in 1874 and to those at Trinidad in 1878. The Colorado Central Company pushed a line up the winding defiles of Clear Creek Canyon, reaching the forks of the stream in September 1872, then laying rails along the northern branch into Black Hawk by the end of the year. This provided Hill with direct rail service for the first time. The Panic of 1873 halted further construction, but when good times returned the Central extended its rails from the forks of Clear Creek to Idaho Springs and Georgetown. And the Denver South Park & Pacific Railway ran a line westward from the capital city to Fairplay and other mining towns in Park County.

The railroad system radiating from Denver had a profound

effect on the minerals industry. As they linked the towns together, the lines gradually reduced the costs of labor, machinery, goods, and services, and they made it possible to ship fuel and ores at far less expense. Hill benefited because the railroads helped lower his reduction costs. This permitted him to process lower grades of mineral and increase his returns to mining companies from an average of 35 percent in 1873 to 65 percent in 1878.[24]

Meanwhile, in 1873, the enterprise faced a dilemma when the refining agreement with Vivian & Sons expired and the Welsh declined to renew the contract unless they received higher compensation. The Vivians' decision left Hill and his associates with three alternatives: they could pay the increase demanded, seek another contractor, or build their own refinery. When they went into business in 1867, Hill and the other founders had not erected a separating plant because of the small smelting capacity of their projected works, a dearth of capital, and the speculative nature of the venture. But those conditions had been obviated by 1873. Two years earlier the enterprise had raised $75,000 for expansion, the firm had tripled the capacity of the Black Hawk plant, and the Alma smelter was now in operation. Since the company had proved itself viable, Hill and his colleagues decided to erect a refinery.[25]

To finance the plant, the enterprise had to augment its capitalization. At the annual meeting in Boston in May 1873, the stockholders voted to increase the equity capital from $275,000 to $500,000. During the next year current owners purchased most of the new shares authorized. Hill took advantage of the opportunity to raise his holdings from 465 to 520 shares, but the treasurer J. Warren Merrill and director J. P. Preston purchased so much of the new issue that they emerged as the largest stockholders.[26]

As in 1867, when Hill persuaded Beeger to emigrate to the United States, the enterprise sought a metallurgist with expert knowledge of the processes to be employed. This time Hill did not turn to Wales, for the man he wanted already lived in Colorado. When he returned to the high country after the annual meeting, he traveled over the mountains to Empire to approach Richard Pearce, the short, redoubtable manager of the Swansea Smelting and Refining Company.[27]

Pearce had had a long career in the minerals industry. He was born in Cornwall in 1837, the son of a mine superintendent who wasted little time in steering his offspring into the same profession. After stints in the workings, Pearce entered the Royal School of Mines, where he had the good fortune to study under the leading experts of the day, including John Percy, who later remembered him as "a remarkable fellow." His formal education completed, Pearce went to work for Williams, Foster & Company, one of the largest smelting firms in Swansea. There he was given responsibility for developing new methods to process copper ores and matte shipped around Cape Horn from Chile. In search of the proper technology, Pearce went to the great continental smelting centers at Mansfeld, Prussia, and Freiberg, Saxony, where he talked with the inventors of the Augustin and Ziervogel processes. He later patented several modifications of the latter method, which he used in a plant he built and operated for Williams, Foster & Company. Although he had a successful career there, as time passed he grew restive because the financial conservatism and resistance of the firm to further innovations made it difficult for him to inaugurate new ideas.

Early in 1871 Pearce received an offer from the newly organized Swansea Smelting and Mining Company, Ltd., to examine its mines at Empire, Colorado, and select a site for a reduction plant. This seemed like an excellent idea, he remembered years later, and he obtained a leave of absence from Williams, Foster to set out for the United States with William Chambers, a director of the enterprise. The trip itself proved exciting, and in future years Pearce enjoyed telling how the great buffalo herds stopped their train for hours. Once in the Rockies Pearce went to Black Hawk where he toured the Boston and Colorado smelter and met Nathaniel P. Hill, perhaps for the first time. Then Pearce and Chambers went on to Empire, a hamlet that had suffered so badly from the process mania that Rossiter Raymond had written that miners had tried everything "from superheated steam to tobacco juice" in a vain endeavor to recover gold from sulfurets. Pearce examined the mines of the Swansea Company, made a side trip to Georgetown, then returned to Black Hawk, where on a dump in Russell Gulch he found samples of uranium ore, a scientific curiosity of little value at the time. This was the first recorded discovery of the metal in the Rockies. On his return to London, he condemned the company's mines but recommended Empire as a smelter location because local pyrites rich in gold could be reduced with silver ores from Georgetown to yield a copper matte that could be shipped to Wales for refining.

With the investigation completed, Pearce resumed his work for Williams, Foster & Company, but only for a short time. Several weeks later the Swansea firm offered him a large salary to erect a smelter at Empire. The proposal was tempting, and the climate provided another inducement because it promised to restore his health, which had been undermined by the dampness of the Welsh coast and the incessant fumes of the smelters. When Williams, Foster & Company did no more than hint at matching the compensation proffered by the Swansea Company, Pearce seized the opportunity to go to the United States. He sailed from Liverpool with his wife and three children in March 1872.

Once in Empire, Pearce purchased a group of buildings along a tributary of South Clear Creek, where there had been an unsuccessful effort to smelt ores by the Swansea process. That summer he directed the installation of crushers, assaying equipment, and furnaces. His firm also hired a group of experienced Cornish workers and sent them to Empire. Everything was ready by fall, and Pearce began operations in October.

Unfortunately for the company, Pearce ran into problems from the outset. He operated the plant only one month in 1872 and only sporadically the following year. Shipping the matte to Wales proved expensive, and refining charges were high, but the chief difficulty was a dearth of ores rich enough in copper to collect the gold and silver. Although many mines along South Clear Creek produced pyrites, Pearce had to compete for them with Welsh smelters, Hill's well-established firm, and the new United States General Mining and Smelting Company, Ltd., which built a plant in the valley that year. And then the Panic of 1873 curtailed mining as well. Pearce tried to increase his supplies of base metal by leasing a mine in Gilpin County, but this did little good. The future of the company was in doubt by the summer of 1873, when Hill arrived in Empire to persuade Pearce to erect a refinery in Black Hawk. Realizing that he might soon be out of work if he stayed with the Swansea firm, he accepted Hill's offer and resigned. It was not long before this short man with a large beard became a fixture at the works of the Boston and Colorado Company. The Swansea enterprise collapsed a few months after he left.[28]

Once in Black Hawk, Pearce wasted no time in putting the refinery into operation. For technology he selected the Augustin and Ziervogel methods that he had used in his work for Williams, Foster & Company. During the summer he directed building construction and supervised the installation of furnaces and other machinery. To staff the new facility, he persuaded Hill to hire some of the "cousin jacks" whom the Swansea Company had brought to Empire. By November all was ready, and Pearce put the refinery into operation. Later that month he turned out his first bars of silver, 99.9 percent pure. The first ingots of gold followed a few weeks later.[29]

Pearce began refining the matte with the Ziervogel process, which removed silver. At the start his "cousin jacks" roasted the matte to drive off excess sulfur and convert the silver compounds into a soluble sulfate. This product was then dissolved in hot water and drawn off, leaving behind a gold-copper residue known in the unscientific jargon of the day as a "bottom." The refiners then ran the silver-bearing solution over copper plates, which precipitated metallic silver. To complete the process, Pearce had the product washed with dilute sulfuric acid and cast into ingots.

Meanwhile, he treated the gold-copper residue by the Augustin process. In this method his workers first added concentrated sulfuric acid to the bottoms to dissolve the copper. Because gold was relatively inert, it was left behind to be melted and poured into ingots. The solution of copper sulfate remaining after both the Ziervogel and Augustin processes was later boiled down and allowed to dry, forming a compound known as bluestone.[30]

The refinery was a technical success, but Pearce and Hill found the Augustin method unsatisfactory because of its high costs. In dissolving the gold-copper bottoms, the process consumed large amounts of sulfuric acid—an expensive commodity in Black Hawk, since it had to be shipped from St. Louis, nine hundred miles away. Hill had to find markets in the East for the bluestone obtained as a by-product. Transportation over such great distances was not only expensive but also inefficient, because the enterprise had to pay partially redundant freight charges. Hill and other members of the firm already considered railroad tariffs much too high.[31]

In hopes of reducing costs, the firm tried a new strategy. In the fall of 1874 Hill transferred Beeger to Boston, where he erected a plant using the Augustin process. Under this arrangement Pearce removed the silver from the matte by the Ziervogel process; then Hill sent the gold-copper bottoms east, where Beeger separated the two metals. The new system produced a small savings in transportation costs, but it still worked badly because assayers in Black Hawk and Boston rarely agreed. This created acrimonious disputes. The situation proved so annoying that Pearce later remembered he was "driven in desperation" to find a method by which all refining could once again be done in the West.

After pondering the problem for a time, he quietly conducted a series of experiments to test an idea he had conceived. Satisfied that the method was commercially feasible, he presented a small sample of gold to Hill, who was incredulous. How had the metal been produced? Pearce gave him the details. Could the process be adapted to large-scale operations? Pearce replied that it could and promised a bar of gold weighing 1,000 ounces the next week. Convinced, Hill telegraphed the refinery in Boston to separate the bottoms on hand and in transit and then phase out operations. For a short time afterward the company used the Bay State works to convert the by-products of the new process into bluestone, then closed the plant permanently in the winter of 1875–76. It was at this time that Beeger left the company's service.[32]

Pearce devised the new process based on the knowledge that copper has a stronger affinity for sulfur than for other metals in bottoms. When he mixed pyrites with the gold-copper mixture, he found that the copper was attracted to the layer of ferrous sulfide that formed above, leaving a lower layer rich in gold. In a sense the copper was stripped from the yellow metal. Three successive strippings yielded a layer of gold 95 percent pure. Oxides of copper were formed as a by-product. Years later Thomas A. Rickard, a mining engineer turned historian, called Pearce's discovery the best technique for separating gold from copper until the advent of electrolytic refining in the late nineteenth century. The method eliminated the Augustin process and permitted the company to reconsolidate its refining operations in Black Hawk.[33]

Never patented, Pearce's discovery remained a company secret for the next thirty-three years. According to Rickard, who was not the most accurate observer, it was the simplicity of the process that baffled the "quidnuncs" who tried to figure it out. But Hiram W. Hixon, a prominent metallurgist, wrote in 1908 that Pearce had been "generally most liberal" in discussing details of the operation. Hixon wondered if the supposedly secret process was "only one of the tricks of the trade."[34]

The evidence tends to support Hixon. In an article published in 1880, Persifor Fraser, a well-known mining engineer, reported that Pearce had divulged the details to his old mentor John Percy. Fraser described the process in general terms and claimed that most metallurgists were familiar with the principle. Shortly after, a rival smelting outfit hired several key employees of the Boston and Colorado firm and acquired the information this way. And Heinrich Roessler, a German scientist, independently rediscovered the method in 1882. In all likelihood what kept the process "secret" was that few other enterprises had need for it.

This controversy notwithstanding, the construction of a refinery created new challenges for Hill and his associates—they had to market the products. Gold was a simple matter. The company sold it to the United States for coinage. Silver presented a more difficult problem, however, because the federal government demonetized the metal in the Mint Act of 1873. The enterprise consequently sold it to manufacturers of jewelry and silverware or else used eastern brokers to sell it on the international market.[35]

Copper had more interesting outlets. At first, the company sold its bluestone to eastern manufacturers who used it to make paving stones. After Pearce developed his secret process, the firm shipped the oxides evolved as a by-product to Boston, where they were converted to bluestone. This, however, proved inefficient. Once Hill and his colleagues closed the Massachusetts works, they sold the oxides to other smelters that reduced them to metallic copper. In later years they sent the oxides to the Standard Oil Company, which used them in refining petroleum from Ohio and Indiana.[36]

Once he put the refinery into operation, Pearce devoted part of his time to improving the efficiency of the smelting furnaces. For some time the company had experienced many problems because it burned wood in fireboxes designed at Swansea, where abundant supplies of coal did little to encourage fuel economy. Combustion in the reverberatories was imperfect because they did not receive enough air to burn all the gases distilled from timber. Pearce solved the problem by boring small holes throughout the length of the roof, admitting more oxygen. Another persistent difficulty lay in the flues connecting the furnaces to their smokestacks. Fires often broke out in the passage, threatening to create another debacle like the blaze that nearly destroyed the plant in 1868. Pearce eliminated this hazard by making an aperture at the base of the chimneys so that air could mix with and cool the hot gases evolved in processing.

Not everyone welcomed such changes, however. Foreman and workers schooled in older methods resisted some innovations, but Pearce managed to persevere because he had the support of Nathaniel P. Hill. The furnace alterations increased the plant's smelting capacity by 30 percent at a saving of one-third the fuel and improved the firm's abilities to process ore. On Hill's recommendation the enterprise gave Pearce a bonus of $2,500 and named him to the official position of metallurgist.[37]

Because of its success the firm drew the attention of the scientific community. As early as 1868 James D. Hague visited the plant as part of his geological exploration of the fortieth parallel. Hill gave him very accurate data regarding operations, and Hague published some in his massive volume, *Mining Industry*, which formed part of Clarence King's study of the region. Then in 1871, in the first issue of the *Transactions of the American Institute of Mining Engineers*, Rossiter Raymond published an article in which he, Anton Eilers, and others discussed Hill's technology. Yet the most detailed of the early studies came in 1876 when Thomas Egleston of the Columbia University School of Mines published in the *Transactions* a lengthy article discussing every aspect of the firm's metallurgy except the secret process. These articles not only disseminated information but also represented some of the first American efforts to build a body of knowledge in a new phase of national endeavor.[38]

As the seventies passed, Hill and his associates continued to expand their smelting capacity in order to process the rising production of mines in the high country. The firm added its fifth and sixth calciners in 1874, as well as a blast furnace to reduce silver-lead ores. The foray into lead-base work was not profitable, however, and the firm quickly abandoned the effort. A year later Hill and his colleagues added a fourth reverberatory smelting unit that brought the ore consumption to forty tons daily. The new furnace had a capacity of twelve tons a day, unusually large for that time, and may have been the result of Pearce's first efforts to increase the size of individual units. Two years later, in 1877, the enterprise erected two more calciners and a fifth smelting furnace, putting the consumption of the works at well over fifty tons daily.[39]

With the capacity of the smelter increasing, Hill opened a string of sampling agencies in Georgetown, Boulder, and Lawson. When mineowners brought in samples of mineral, Hill's men crushed the rock until they obtained a representative sample; then each party assayed its half. If the miners agreed with the determination and price offered by the smelting company, they signed a contract, and the agency had the ores shipped from the mine to Black Hawk. If the two could not reach an agreement, the mineowners paid Hill's company a small fee and opened negotiations with someone else. The sampling agencies did not end disputes between Hill and his shippers, but they speeded operations and aided the movement of ore to Black Hawk.[40]

During this time Hill's sources of ore remained largely unchanged. He continued to draw the bulk of his mineral from mines in Gilpin, Clear Creek, and Boulder counties. The Black Hawk works processed from 20 to 25 percent of their production. The Alma smelter, meanwhile, with a near monopoly, reduced about 90 percent of Park County's output. Hill purchased ores from other mining districts, but they made up a relatively small part of his shipments.

Even though Hill's markets remained relatively constant, silver gradually superseded gold as the primary product. In the early seventies the company recovered about equal amounts of the two metals, but after 1873 the value of the silver production outstripped that of gold by two to one. This shift was the result of the rising production of silver-bearing ores in Clear Creek and Park counties. Hill's shipments of copper also increased, but their value never exceeded 5 percent of total production. The importance of copper, however, transcended its low monetary value because the metal was essential in recovering gold and silver. Neither Hill nor Pearce nor other members of the corporation could know it during the seventies, but thirty years later a dearth of this metal would have dire consequences for the enterprise.[41]

In spite of Hill's importance to the economic development of Colorado, there was one deleterious aspect to his success—the Black Hawk smelter was a severe polluter of the clear Rocky Mountain air. What made the situation even worse was the location of the plant and the town in the narrow, high-walled canyon carved by North Clear Creek. Roasting, smelting, and refining all produced vast quantities of sulfur dioxide and other noxious gases that filled the skies above the town. Yet this was regarded as a symbol of industrial progress during the Gilded Age, and the company's stationery proudly depicted the plant shrouded by billowing clouds of smoke.

The pollution was so bad, however, that travelers had to run a

Silver bars produced at Hill's smelter in Black Hawk. Richard Pearce stands at the extreme left, William A. Abbe in the doorway. Pearce's former employer, the Swansea Smelting and Mining Company, brought the four smelter workers from Wales to the United States. L. McLean, photographer. Western Historical Collections, University of Colorado Libraries, Boulder.

gauntlet in getting into town. A newspaper correspondent from Michigan wrote in 1878 that Hill's smokestacks emitted "volumes of blackness" that filled "the atmosphere with coal dust and darkness." A staff correspondent for the *Engineering and Mining Journal* was even more eloquent. He reported that as his train steamed tortuously up North Clear Creek, he and his fellow passengers were engulfed by "villainous vapors . . . choking gases . . . and cough-compelling odors" that they had to endure for more than a quarter mile before reaching the depot in

Black Hawk. This correspondent also remarked that pollution had given the town a bad reputation it might always retain unless Professor Hill moved the works to another locality. This was exactly what Hill was planning to do, though not because his plant had a harmful effect on Black Hawk's public image.[42]

Until the late 1870s Hill had smelted ores drawn largely from the valley of Clear Creek, but late in the decade he began to look farther afield. To some extent this evinced a natural desire for expansion, but it also reflected a need for the copper essential in the Swansea process and the hurly-burly of western railroad construction that made large interregional ore shipments possible for the first time. An important break from tradition came in 1877 when William Andrews Clark, then on the eve of his turbulent career in Montana, arrived in Black Hawk to talk with Hill, Pearce, and Wolcott. Clark had shipped some of his mineral from Butte to smelters near Salt Lake City, but the cost of transportation and treatment had left him with little profit. This was hardly satisfactory. He wanted to know if Hill's works could do better.

Hill was interested, since he was in business for profit. When Pearce assayed Clark's ore samples, he found they held from 15 to 20 percent copper and as much as 250 ounces of silver per ton. The high silver content made the ores attractive, but the high percentage of copper made them particularly alluring because Hill and Pearce had noticed the quantity of copper in sulfurets declining. Hill granted Clark a favorable smelting rate, and the man from Butte sent $278,000 worth of ore to Black Hawk the next year. This represented about 13 percent of Hill's output and marked the first time he had processed ores mined outside Colorado.[43]

But Clark's shipments and the rising production of Colorado created new challenges. The steady expansion of the plant made the original four-acre site in Gregory Gulch very cramped. To make room for the furnaces erected in 1877, Hill had had to level off and build over the slag dumps. Pearce had also increased the size of the roasting mounds and lengthened the burning time, since he found it impossible to enlarge the roasting yard. The sulfurous pollution, however, continued without pause, causing some concern among the company's staff. Pearce, for one, had left Swansea partly because smelter fumes had impaired his health. Hill also saw the cost of fuel rising because deforestation around Black Hawk required him to purchase timber from ever more distant forests. Success had created its own problems and restraints on continued growth—and possibly even on continued prosperity. All this prompted Hill and his colleagues to give some thought to the idea of relocating their operation.[44]

As rumors of an impending move began to circulate, Hill saw the company confronted by an unexpected challenge. Late in 1877 Secretary of the Interior Carl Schurz induced the Justice Department to prepare a damage suit against the firm on the grounds that it had removed trees from public lands without paying the federal government. When Hill learned of this, much to his consternation, he wrote Senator Henry M. Teller that if the action were successfully prosecuted, it would put an end to the company's operations at Black Hawk. Hoping to avert a trial, Hill asked Teller to do him a "great favor" by using his "influence to dispose of the case without litigation." When Hill failed to receive a favorable reply, he went to Washington to discuss the matter with both Teller and Schurz, but his conversations failed to produce the desired result.[45]

Official proceedings against the company began in November when Westbrook S. Decker, the United States attorney in Denver, filed suit in circuit court. He asked for $100,000 in damages, claiming that from July 1875 to September 1877 the Boston and Colorado Company had failed to pay for fifty thousand cords of wood that it had taken from the public domain. As a defense attorney Hill engaged Henry Wolcott's younger brother Edward, a recent graduate of Harvard Law School. Through him the enterprise denied the charge, contending that it had purchased the timber from individuals who had taken it "from the mineral lands of the public domain, as they had a right to do." Displaying his skills as a trial lawyer, Wolcott obtained a jury favorable to the interests of the smelting company. This allayed much of Hill's anxiety over the trial, but the case dragged on for another year before the jurors rendered a verdict supporting the firm. By then Hill and his associates were phasing out their operations at Black Hawk.[46]

The federal suit brought the question of relocation to a climax. Hill's fuel costs were already rising steadily because of local deforestation, and he knew that if Decker won the case prices would climb still higher and might eliminate profitable operations altogether. But beyond the federal suit, Hill and his colleagues knew that the cramped site in Black Hawk prevented them from increasing capacity any further—at a time when Colorado's production was on the rise. As their future prospects in the mountains looked ever more bleak, Hill and his associates looked with greater favor on the idea of consolidating operations on the plains.

Rail transportation now made the difference in the decision to remain at Black Hawk or relocate in the "valley." Unlike the situation in 1867 when they went into business, Hill and his associates saw that the roads built in the past few years would permit them to draw on many mining districts whose production could roll downhill to the smelting furnaces. The lines that radiated from Denver would also permit the plant to tap ores mined hundreds of miles away, outside Colorado. Such a move would also enable Hill to switch his primary fuel from wood to coal, as he had long wanted to do. Pearce's experiments had found the coal mined at Trinidad and Canon City an excellent source of energy, but one impossible to use at Black Hawk because it was too expensive to haul up Clear Creek Canyon. Beyond these considerations, Hill knew that prices and labor costs were lower on the plains than in the mountains and that he could acquire a tract of land large enough to meet the requirements of future expansion. And so in the winter of 1877 Hill and his associates decided to consolidate all their operations near Denver, the hub of Colorado's railroad network.[47]

Operations in the mountains continued for one more year as Hill went ahead with plans for the new complex. During the summer of 1878 Henry Williams closed the Alma plant and left for Montana to investigate the potential for still another smelter. Wolcott continued to reduce ores at Black Hawk, but as the year passed he diverted more and more shipments to the new works. It was not until December that the smelter reduced its last ton of ore. Yet the plant was not torn down or sold because, once the new complex opened, Hill converted the older works into a sampling agency in order to maintain his strong position in the ore markets of Gilpin County.

Yet the closing of the Black Hawk and Alma smelters ended a notable chapter in Colorado's history. Since 1868 Hill's firm had been the only reduction enterprise in the Rockies and one of few in the United States to attain both technical and financial success in smelting gold and silver ores. Production had grown from $271,000 in 1868 to $2,260,000 in 1878, a compound annual growth rate of 23.6 percent. This output accounted for 25 to 37 percent of the value removed from the mines in the high country during that time. When Hill and his colleagues opened their new works on January 1, 1879, their enterprise had a promising future, but in the year just past new developments in the central Rockies had altered the course of the mining industry—and with it the role of the Boston and Colorado Smelting Company.[48]

Chapter 3

The Struggle for Survival

IN THE SUMMER OF 1864 A BAND OF TREASURE SEEKERS DISEN-
chanted with their declining prospects at Empire followed the
trace of South Clear Creek westward to the foot of Loveland
Pass, the gateway to central Colorado. From there they crossed
the divide to the headwaters of the Snake River, where they
prospected for mineral on the rocky slopes of the high peaks. No
one knows whether the fortune hunters were searching for gold
or silver, but on Glacier Mountain they located the gray outcropping of a deposit of silver-bearing lead ore. This particular
claim proved insignificant, but its discovery—and the knowledge of what to look for—spurred prospectors on both sides of
Loveland Pass. The silver boom was on.[1]

From the outset miners realized they would have to employ a
smelting process to recover the precious and base metals. The
ores occurred in the form of galena, a lead sulfide holding silver
compounds, and milling did little to "beneficiate" the mineral.
The question was this: What was the correct smelting method?
Scotch hearths? Reverberatory furnaces? Or something else?
Many thought the answer was the hearth or the reverberatory
because they had been used successfully in the lead regions

of Missouri, Iowa, and Wisconsin. Minerals there, however, held no silver, the valuable component of Rocky Mountain ores. Yet this dissimilarity made little difference to the men who inaugurated silver-lead smelting in Colorado.

Even though the Snake River district lay in extremely isolated country, fortune hunters hearing news of silver made the arduous trek there, and the hamlet of Saints John sprang to life as the hub of the minerals industry. Most of the newcomers had neither the capital nor the inclination to open mines and ship ores. What they wanted to do was file claims on promising properties that they intended to sell to eastern investors, a scheme that worked well for a time, as it did in every mining region.[2]

The most important mine in the district was the Comstock lode, shrewdly named for, though far less rich than, the renowned silver mines of Virginia City, Nevada. Still, assays of the Colorado property suggested that the deposit held unusual riches, and the Boston Silver Mining Association purchased the lode in 1865. To handle development the company hired John Collom, a Cornish mining engineer with long experience in Europe and America. He opened the mine, erected ore-dressing machinery to concentrate the mineral before reduction, and built a smelting plant with six Scotch hearths to reduce ores to silver-lead bullion. He put the works into operation in the summer of 1867 and produced a single ton of bullion that he shipped east to Newark, New Jersey. There the product was refined by the firm of Edward Balbach & Son, which was soon to play a signal role in Western smelting. Heavy snows forced Collom to suspend operations during the winter, but in the spring he smelted more Comstock ores into nine tons of bullion that he again shipped to the Balbach refinery for separation into silver and lead.

This success proved illusory, for Collom soon realized that Scotch hearths were not adequate for smelting silver-lead minerals. The furnaces lent themselves to the wilderness because they were inexpensive to construct and operate, but they had one crucial limitation—they required an ore charge at least 60 percent lead to collect the silver. Collom rarely mined rock that

The smelter at Saints John built by the Boston Silver Mining Association. Compare the simplicity of this small, isolated plant of the early 1870s with the large, integrated Globe and Durango smelters of 1900. Colorado Historical Society, Denver.

rich, and his ore-dressing machinery could not increase the concentration of base metal much past 50 percent. Not surprisingly, he lost large amounts of silver in the process of reduction. This was the key problem, but the association also had to pay shipping and refining charges that amounted to eighty-three dollars per ton of bullion. In such circumstances Collom could reduce only the richest ores that his miners took from the Comstock. By the end of 1868 huge quantities of mineral worth as much as $200 a ton were piling up on the dump because the company could not bear the cost of smelting, shipping, and refining.[3]

Collom thought he knew a solution. In 1868 he told James D. Hague that the only feasible way to reduce silver-lead ores was to use a blast furnace. Hague, who was familiar with the process because of his own education in European mining schools, made a note of the idea, but Collom resolved to put his belief into operation. Early in 1869 he built a smelting unit similar to one he had seen in France, but he did not have the furnace in successful operation when Hague visited Saints John later that year. During the seventies the association used a blast furnace to process its ores, but whether it was erected by Collom, who had long since left the enterprise, remains unknown. Nevertheless, Collom may have been the first person in Colorado to adapt, build, and operate a blast furnace.[4]

In the meantime, fortune hunters spurred by the news of silver on Snake River sought out the gray black outcroppings of galena in the valley of South Clear Creek. These men centered their efforts of the precipitous mountains that rimmed the skies above Georgetown, an early gold camp whose fortunes were on the wane. On September 14, 1864, prospectors spied the surface markings of a silver-lead deposit they named the Belmont lode. The cry of silver leaped across the region, treasure seekers rushed to the site, and within months a lucky few had filed claim to the Baker, John Brown, Coin, Equator, and Terrible lodes situated near the village.[5]

The first ores from Georgetown dazzled the industry. Some assays revealed minerals worth upward of $3,000 a ton in silver and lead, a reflection of the extensive secondary enrichment. The rock beneath the surface held far less wealth, but this mattered little to early enthusiasts, scrupulous or unscrupulous. Prospectors sold claims for fantastic prices, and mining companies advertised their own securities as gilt-edged investments in "inexhaustible" resources. Such proclamations snared the unwary and mesmerized even the knowledgeable, who spent vast sums in mine development and smelter construction only to find complex, lower-grade ores that the reduction plants were unable to process with any degree of utility.[6]

Like Lyon, Collom, and others, the first smeltermen in Georgetown turned to furnaces used in the lead regions of Missouri, Iowa, and Wisconsin. The Argentine Silver Mining and Exploring Company installed a Scotch hearth near camp and experimented with a refining process recently invented at Portland, Maine. The Bohemia Smelting Company, which owned a "strong galena vein," erected a reverberatory unit that supposedly worked "to the satisfaction of the owners," though not for long. Another smelterman knew so little about reduction that he had to hire a former slave, who had worked in the lead region of Virginia, to operate his furnace. And a highly touted Professor Dibben met with utter failure after building a unit for which the community held high hopes. Such early smeltermen had little knowledge of metallurgy and no practical experience. Rossiter Raymond remarked that they "seem to have believed that nothing was necessary to accomplish a success . . . except common sense, fire-brick, ore, and wood."[7]

The first enterpise to achieve even a modicum of success was the Georgetown Silver Smelting Company. In the spring of 1867 the New Yorkers who had organized the firm hired John T. Herrick, a mining engineer, to erect a reverberatory furnace for smelting ores and a cupelling unit for refining bullion. Herrick built the plant and conducted a short "campaign" that summer. Technical difficulties beset the company, however, and in the fall Herrick leased the plant to the firm of Schirmer and Bruckner, two mining engineers educated in Europe.

Schirmer and Bruckner invested $1,500 to remodel the works and install an experimental roasting furnace Bruckner had designed. Then they purchased ores from local mines, put the plant into operation, and recovered as much as 95 percent of the silver in certain classes of mineral, a very credible showing for the time. The rate of recovery dropped sharply, however, when the partners smelted refractory rock from the justly named Terrible lode, Georgetown's showcase property. The season of the year also plagued the enterprise. Winter was a difficult time for smelting ores in the mountains (and would hamper many entrepreneurs in the future). Schirmer and Bruckner could not obtain enough ores, fuel, and fluxing materials to keep running. Water froze in the pipes, and the firm had trouble with other machinery. When these problems could not be overcome, the partnership failed. But in spite of its demise, the firm left a significant legacy. Bruckner's roasting cylinder became the prototype of a furnace used throughout the smelting industry until the twentieth century.[8]

Early in 1868 the plant reverted to Herrick's company, which resumed operations using the improvements made by the former lessees. This time the firm had better luck. By August Herrick reduced about ninety tons of ore at an average cost of twenty-eight dollars a ton. James D. Hague thought the figure excessive, but Herrick insisted that shipments holding less than one hundred ounces of silver and 30 percent galena could not be mined and smelted profitably. His price schedule tended to bear this out. He paid mineowners only 45 percent of assay value for ores meeting the minimum specifications. Payment rose for richer mineral. Yet Herrick also employed a strategem that augmented his income at the expense of mining companies. He paid his shippers in greenbacks, fiat money issued during the Civil War and worth only about two-thirds the face value of hard money, known as coin. This practice, used by many early smelting firms, infuriated mineowners and added another element of discord to the tepid relationship that prevailed between producers and processors.[9]

Despite a measure of success, Herrick still had to contend with technical problems. Like James E. Lyon and Nathaniel P. Hill, he had built his furnace with the low-quality firebrick manufactured at Golden—material that permitted bullion to leak through the masonry. But that was a relatively minor problem. The real impasse was that the reverberatory furnace was unsuited for smelting the ores. The lead content of Georgetown's minerals fell far short of the 60 percent required to collect the silver. Herrick was apparently unaware of the limitation, for he told Hague he could recover 90 percent of the metal content with a better furnace, and he hoped to persuade his financial supporters to appropriate funds for one. But, as events proved, he was unsuccessful. The Georgetown firm ceased operations in 1869 for what Hague termed "financial embarrassments," the euphemism of the day for business failure.[10]

The Brown Silver Mining Company was another early firm that fell into a technological trap. Organized by Philadelphians in 1867, the enterprise purchased the John Brown, Coin, and other properties "for legitimate and profitable mining, and not for stock speculation," according to the first annual report. To manage operations in Colorado, the company hired Joseph W.

Watson, a knowledgeable mining man, part-owner, and former associate of Nathaniel P. Hill. Watson opened the properties and sold ores, some worth more than $400 a ton, to both the Georgetown and the Bohemia smelters. Returns were great enough for the company to pay dividends every month.

Watson and his colleagues were so encouraged by their success and so eager to avoid reduction charges that they decided to build their own smelter. While there is no evidence that Watson knew anything about high-heat processing, he directed the construction of a small reverberatory furnace as well as roasting and refining units. Shortly before operations commenced in the summer of 1868, he told Hague the plant would cost from $80,000 to $100,000, a figure the shrewd Bostonian privately thought excessive.[11]

Watson opened the smelter that fall, but he found that the galena mined from the Coin and John Brown lodes had too little lead for his furnace to use in recovering silver. Large amounts of silver escaped in the slag. And, adding to the company's woes, the mineral contained high percentages of zinc, which interfered with reduction. Watson did not give up easily. He purchased wagonloads of galena mined near Central City, had the ores shipped to Georgetown, and mixed them with Coin and John Brown rock. Yet the amount of lead still remained too low to capture enough silver to permit profitable operations. When this expedient failed, he had pigs of lead shipped from Chicago, despite freight charges amounting to $120 per ton. These measures enabled the Brown firm to remain in business until 1870, but by then the company was $80,000 in debt. Watson found himself besieged by creditors and fled to Utah to recoup his fortunes in the burgeoning silver-lead industry of the Salt Lake Valley. He placed the Brown Company in charge of an agent, who sold the mines in 1872. The smelter, however, was a total loss.[12]

Watson's failure ended the first era of silver-lead smelting in Colorado. Since James E. Lyon's first effort back in 1865, many industrialists had turned to hearths and reverberatories because they had proved successful in the Missouri and Mississippi valleys, where the ores held high percentages of lead and little or no silver, zinc, copper, or other metals. But in the Rockies the situation was the opposite. Here, the primary value of the lead was as a collector of silver. As they re-created the society they had known elsewhere, it was natural for the early miners to rely on traditional methods of reduction. This was unfortunate. Many investors might have avoided financial ruin if they had consulted experienced metallurgists; but these were rare, and the specter of precious metals clouded the judgment of people who might otherwise have been more prudent. Only after a period of trial and error brought disaster to Lyon, Herrick, Watson, and countless others did the use of impractical methods come to a dismal end.

What happened in the Rocky Mountains, however, was only a reflection of what was taking place across the West. As early as 1860, prospectors had located silver-lead ores in Nevada, Utah, Montana, and other territories. The first entrepreneurs to exploit the deposits erected hearths and reverberatories that produced the same dismal results as those in Colorado. So it was not just investors with mines in the high country, but investors throughout the nation who demanded a technology that could reduce silver-lead ores and recover the precious metal. The resolution of the problem proved difficult at first because the mines lay in an undeveloped region of an underdeveloped country. Yet, even as the industry stumbled over hearths and reverberatories, a handful of metallurgists were introducing the proper technology, a method used for years in central Europe, where the theory and practice were well advanced.

It was in the middle years of the century that Americans first went abroad in large numbers to study mining and metallurgy. The Royal School of Mines, on Jermyn Street in London, drew some students but not many, since mining engineering and metallurgy lacked prestige in Britain. Only a few went to the Ecole des Mines in Paris, because the French government severly restricted the admission of foreigners. The Mining Academy of Clausthal also attracted Americans because of its excellent reputation. But by far the most famous school was the Königliche Sächsische Bergakademie at Freiberg, Saxony, a few miles up the valley of the Mulde River from Dresden. Most Americans who studied mining engineering abroad came here to work with the leading men in the profession and gain access to the state-operated smelting works. It was no coincidence that graduates of the Bergakademie pioneered the use of new technology in the American silver-lead industry. Yet, regardless of mining school, no one thought a European education complete without a summer tour of the metals centers at Schemnitz in Hungary and Przbram in Bohemia.

Besides the Americans who went abroad to study and return with the latest developments in mining and metallurgy, there were German graduates who emigrated to the United States to enter the minerals industry. Some came on their own initiative. Others arrived because American companies had sought them out, much as Nathaniel P. Hill had sought out Hermann Beeger in 1867. Still another group came over because mining engineering firms like the partnership of Justus Adelberg and Rossiter Raymond wanted to hire the best-educated men.[13]

It was this group of men, all trained in Europe, who introduced the blast furnace—the proper technology for smelting silver-lead ores. When the method appeared in the United States in the late 1860s, two models were in widespread use abroad. One was the Raschette furnace, which rose from a rectangular base into a form that resembled an inverted, truncated pyramid. The other was the Piltz furnace, which rose from a polygonal or circular base into a shape that resembled an inverted, truncated cone. The base of the Raschette model measured three feet by five feet, while the base of the Piltz unit was about three and one-half feet in diameter. The interior of each was lined with firebrick. Once the furnace was in operation, smelter workers added ores, fuel, and flux in a specific order through an opening near the top. Blowing engines, which gave the furnace its name, pumped air into the interior through apertures, known as tuyeres, near the bottom; they were not protected by water jackets when the furnace appeared in the United States. This was an American innovation, although devised by a German-educated metallurgist.[14]

Fuel served a dual purpose in the blast furnace. Unlike the

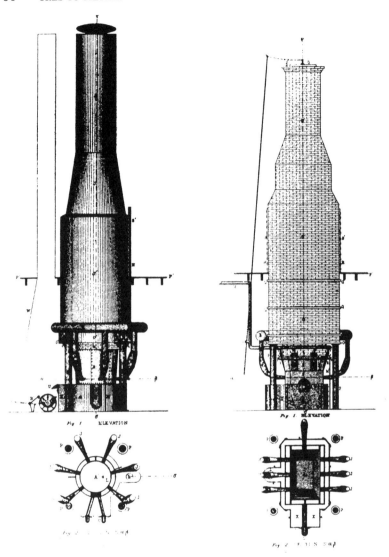

Fig. 3. Piltz furnace. Drawing by Antony Guyard. Reproduced from Emmons, *Leadville*.

Fig. 4. Raschette furnace. Drawing by Antony Guyard. Reproduced from Emmons, *Leadville*.

hearth or reverberatory, which burned wood or charcoal in fireboxes separate from the ore charge, the blast furnace consumed charcoal or coke to produce greater heat and also to provide the carbon essential in the chemical reactions that occurred in the reduction process. Like their counterparts in the iron industry, silver-lead smelters preferred coke, but there were many problems that attended its use in the early days of western smelting. A coal low in sulfur and low in ash was a prerequisite for a high-grade metallurgical fuel, but the first silver-lead furnaces that appeared in the West were hundreds of miles from the only source of adequate coking coal, the town of Connellsville in the mountains of western Pennsylvania. The cost of transportation plus the working of supply and demand pushed prices in isolated mining camps to as much as sixty dollars a ton. Little wonder that early metallurgists hoped to find western coals comparable to the Connellsville product. In the interim, smelters turned to charcoal made from nut pine, fir, and spruce, but it was a poor substitute. Edward D. Peters, Jr., an early Colorado smelterman, called it "carbonized sawdust." Yet it served its purpose until coke came into abundance.[15]

What happened in the smelting process depended on the composition of the ore charge. Reduction to bullion was relatively simple if the mineral was a carbonate, which in the loose parlance of the day meant an oxide, sulfate, or carbonate of lead. Once the furnace had heated the ores to a molten state, the charcoal or coke reduced them to silver and lead, which formed a dark layer of bullion below the slag. Sulfur and carbon dioxide gases evolved in the process escaped into the atmosphere.

Galena like that mined near Saints John and Georgetown proved far more difficult to reduce because nature had done little to convert the mineral into "carbonates." Smelters had to roast these ores to drive off some of the sulfur and convert the remainder to sulfates, but neither reaction eliminated all the sulfide. Thus metallurgists had to add a fluxing agent to the charge; iron ore was the most popular because it easily displaced the sulfur, freeing the lead to collect silver. Adding iron, however, complicated the smelting process because it formed a thin layer of silver-bearing matte that coalesced between the lead bullion and the slag.[16]

Yet smelting was rarely that simple. Most ores held small amounts of zinc, manganese, copper, iron, gold, and other metals that further complicated operations. Zinc compounds formed accretions on furnace walls, impeding the movement of bullion, matte, and slag. Manganese, iron, and limestone could all be used as fluxes, and metallurgists sometimes used them together. Copper increased the quantity of matte and coalesced with iron to capture silver and gold that might otherwise have found their way into the lead-base bullion. Western ores also held traces of arsenic, cadmium, cobalt, thallium, vanadium, and other metals, although they did not become important smelter products until the twentieth century.[17]

Despite the complexities of operation, the capabilities of the blast furnace suited the needs of the mining industry. It could reduce ores holding small quantities of silver and lead as well as varying amounts of other substances. The conspicuous success enjoyed by several early firms popularized the method at the same time that hearths and reverberatories were passing into oblivion. Many firms built blast furnaces over the next few years, but only a few survived for any length of time. There were many reasons for this—inadequate ore supplies, bad transportation, dearth of capital—but often the key reason for failure was the lack of a skilled metallurgist familiar with the process. As events showed, the presence of an experienced person was often the sine qua non for profitable operation. For this reason an important figure in the industry's early development was the engineer-entrepreneur, who could solve interrelated technical and financial problems. Such men were usually high in the managerial pyramid, and a few emerged as leading industrialists of the Gilded Age.[18]

The blast furnace made its first appearance in the West during the late 1860s. Almost simultaneously, mining companies at Argenta, Montana, and Oreana, Nevada, erected units to process nearby ore deposits. Although the plants were short-lived, they were followed by others in the dusty, windswept town of Eureka, Nevada. It was in this desiccated country that the Richmond and Eureka mining companies had such success in working the ores of Ruby Hill that Eureka became the cradle of the silver-lead industry in the trans-Mississippi West. Outside Eureka, smelting firms made little progress in the wastes of Nevada, but the industry took strong root in Utah, where prospectors had located silver-lead ores in the Salt Lake Valley.[19]

Smelters appeared in Colorado about the same time they emerged in Nevada and Utah. In 1869 John Collom erected a furnace at Saints John, and two years later William West built one at Black Hawk. But neither man had much success. Collom soon left his company, and West sold out to Hill, who abandoned the furnace after a brief trial. More significant was Edward D. Peters, Jr., who in 1872 built a plant at Dudley, a new mining town in South Park.[20]

Industrial development in this region had followed the episodic nature characteristic of western mining. Once the initial boom collapsed, depression stalked the country until prospectors found silver-bearing ores on the slopes of Mount Lincoln and Mount Bross. The influx of miners and infusion of capital that followed revitalized the district. Hill and his colleagues built the Alma smelter to process pyrites, but at the same time another group decided to erect works to reduce silver-lead ores. The chief entrepreneur was Judson H. Dudley, an old fifty-niner who was now the chief stockholder in the Moose Mining Company, which owned several claims in the district. In the winter of 1871 he provided the impetus for the formation of the Mount Lincoln Smelting Works Company.[21]

Rather than build a hearth or reverberatory, which they must have known would be a waste of time and money, Dudley and his associates turned to Peters to erect a blast furnace. How Peters came to Dudley's attention remains unknown, but he was young—just twenty-two years of age—and on the threshold of a great career in metallurgy. A native of Massachusetts, he had studied at the Bergakademie, where he had learned about blast furnaces, and he had just finished his first job building a silver mill in Nederland, the supply center for Caribou. He was interested in new employment.[22]

Peters began the plant that summer. While construction crews worked on the buildings, he carefully directed the installation of a blast furnace based upon the Piltz—or circular—model that he had studied in Europe. At the same time the firm purchased silver-lead ores from local producers, Dudley's Moose Mine shipping more than three hundred tons of mineral worth some $350 each. When Peters and his colleagues "blew in" the smelting unit in December, everyone involved in the project held high hopes of success.

Operations went badly, however. With coke unavailable at any price, Peters had to burn charcoal produced in beehive kilns from local fir and spruce trees. The fuel, however, was low in quality and expensive to manufacture. Peters also had to purchase his iron flux at a loss; and large quantities of spar in his ores added to the difficulties of processing. Yet with careful work Peters reduced his mineral to bullion in the course of "campaigns"—the length of time the furnace remained in blast—averaging about three weeks. He shipped the bullion to Germany for refining. In all, the costs of reduction amounted to the astronomical figure of sixty-four dollars a ton.

Despite his technical success in the face of adverse conditions, Peters and his colleagues found themselves unable to overcome a crucial problem—a dearth of high-grade ores that could bear the cost of reduction. Even though Peters offered from twenty to fifty dollars more per ton than eastern or European processors, he could not prevent mineowners from shipping ores out of the district, nor could he stimulate greater production of high-grade minerals. After nine months of effort Peters and his colleagues concluded that the Lincoln-Bross mines would never provide enough ores to permit the smelter to operate profitably.

By August 1873, Peters and his associates decided to abandon the blast furnaces in favor of the Swansea process. The next month he erected a reverberatory furnace and entered the local ore market to buy a good supply of pyrites. While he was engaged in the conversion, Dudley's group sold a half-interest in the company to two investors in Philadelphia. Once operations resumed, Peters's reduction costs fell sharply from sixty-four to twenty-nine dollars a ton—a decline of more than 50 percent. Using wood for fuel and eliminating the iron flux purchased at a loss accounted for much of the decrease. With the plant on what seemed to be a better footing, Dudley and his colleagues sold their remaining half-interest in the enterprise to investors in Dubuque, Iowa.

In spite of the changes, the Mount Lincoln company still found itself beset with problems. Railroads charged high freight on matte, while Peters and his German refiner never seemed to agree on assay values. Yet, even worse, by the end of 1873 Hill's Alma works had nearly monopolized the district's production of sulfurets. As Peters found himself unable to break Hill's domination of the market, he could not maintain steady operations. The situation deteriorated during the winter, and Peters had to cease operations in January 1874. Dudley and his associates then filed suit against the new owners for nonpayment of $30,000, but this made little difference, for the enterprise soon went out of business.[23]

Despite the failure of the Mount Lincoln company, Peters went on to a successful career in metallurgy. After leaving the firm, he returned to Boston, where he wrote an article discussing the problems of using blast furnaces in isolated regions. This study, published in the *Transactions of the American Institute of Mining Engineers*, was one of the earliest contributions by an American-born metallurgist to the art and science of smelting silver-bearing lead ores; it was also the initial firsthand account of the use of blast furnaces in Colorado, the most important center of the industry less than a decade later. Peters, however, was discouraged by the failure of the Mount Lincoln enterprise. For a time he took up the study of medicine, but once he received his degree he returned to his original profession. In the next four decades he helped establish copper smelters throughout the world and finally was appointed professor of metallurgy at Harvard.[24]

An even more ambitious effort at silver-lead smelting came

in 1873 when a group of British investors organized the Hall Valley Silver-Lead Mining and Smelting Company, Ltd., to acquire mines and build a smelter in that isolated hollow of Park County. The surface ores were rich—galena averaging thirty ounces and pyrites three hundred ounces of silver per ton. To process the mineral, the firm erected a smelter "without regard to cost," according to Rossiter Raymond. The plant consisted of three blast furnaces ostensibly built on the Piltz model, though they bore a greater resemblance to units employed in iron smelting. The firm used charcoal for fuel and bog iron for flux.[25]

Despite the hopes of the founders, the enterprise ran into technological problems. The modified Piltz units may have been well adapted for reducing iron ores, but because of their height they could not maintain the proper temperatures for smelting a mixture of galena and pyrites. The recovery of bullion and matte disappointed everyone concerned, but the company did not give up easily. Thinking that an increase in flux might augment the return, the enterprise bought limestone in Fairplay and iron-bearing slag from Hill's works at Alma. When they were added to the ore charges, they did increase the amount of bullion and matte recovered, but not enough to permit the firm to operate profitably. Company metallurgists tried other remedies, but to no avail. The furnace structure prevented remunerative reduction, and operations eventually ceased. In spite of its failure, however, the Hall Valley smelter left an important legacy in the technical literature of the day. Joseph L. Jernegan, a young metallurgist later drowned in crossing the Colorado River, wrote two articles published by the American Institute of Mining Engineers. He discussed the firm's difficulties, noted the "disadvantageous shape" of furnaces designed for iron smelting, but saw "no difficulty" for future smelters that employed a better furnace design.[26]

Besides the Mount Lincoln and Hall Valley firms, there were other companies that used blast furnaces during the 1870s. The St. Laurence Mining Company, the Lincoln City Lead Works, and the Golden Smelting Company all made brief appearances in the industry, but they met with little success and disappeared leaving little imprint. So it was in southwestern Colorado that several firms erected smelters, but their operations were sporadic and they remained marginal enterprises.[27]

Unlike Hill's immediate success with the Swansea process, the first efforts to use blast furnaces nearly all came to failure. The smeltermen faced heavy odds. High costs were normal, fuel was poor, ore supplies were inadequate, and metallurgists were scarce. The general lack of industrial development throughout the territory magnified every problem. Many overly optimistic investors suffered severe financial reverses.

Because the early smelters proved unequal to the task, mineowners had to look outside Colorado for a market. They shipped some ores to Europe, but most went only as far as Omaha or St. Louis, where two firms with ample capital, low costs, and superior metallurgical technique provided a ready market. Yet the costs of shipping, smelting, and refining still prevented all except high-grade ores from being mined in the high country.

Eastern firms had several advantages over their counterparts in the Rockies. The more advanced development of railroad systems and natural resources gave eastern smelters greater access to metallurgical fuels and enabled them to draw upon a wide variety of ores instead of relying on the production of one district or even one mine. In fact, their location far from western mining camps was a stroke of good fortune because it compelled firms to tap many ore markets and shielded entrepreneurs from the temptation to sink investment capital into mining properties. Although Rossiter Raymond warned in 1871 that the success of distant smelters depended upon the goodwill of railroad companies, ores provided attractive freight, particularly for financially troubled lines. Many railroad tycoons later went into the smelting industry as a sideline.[28]

One important eastern firm came into existence on October 15, 1870, when eight men gathered in Nebraska to organize the Omaha Smelting and Refining Company "to deal in, reduce, smelt, separate, and refine metals, ores, and minerals." The group consisted of C. H. Downs, W. W. Lowe, C. W. Meade,

A. L. King, and John A. Hoorbach, all respected businessmen in the city of Omaha, and Leopold Balbach and Henry Deiffenbach of the small but well-known Balbach smelting firm of Newark, New Jersey. They set the capitalization at $60,000, a small sum for a smelting enterprise even at this early date. The directors elected Downs president, Lowe treasurer, and King secretary, but the actual correspondence fell to Edward W. Nash, who would be one of the most important figures in the industry's development.[29]

Once organized, the company employed expert metallurgists —in this case Charles and Leopold Balbach—to build and operate the plant it established in 1871 on an industrial tract at the foot of Capital Avenue in Omaha. The concern ran its business, however, through offices at 195 Farnham Street. The company drew its ore supplies largely from mines situated in country served by the Union Pacific Railroad. Since there was no bridge over the Missouri River from Omaha, the smelter literally sat at the end of the line.

Through their close connection with officials of the Union Pacific Railroad, the Omaha smeltermen had an important competitive advantage. This was particularly true during the 1870s, when the road was in a precarious financial condition and wanted to assure itself of traffic from western mines. E. P. Vining, the freight manager, wrote one shipper that the Union Pacific "was anxious to encourage" the smelting company. Since the road had a vested interest in the firm's prosperity, it gave the men of Omaha rebates, which they split with mineowners. This was an effective practice in the competitive markets of Colorado, Utah, and Nevada. In future years Sidney Dillon, president of the railroad, served as a director of the smelting enterprise.[30]

The Omaha Company prospered almost from its inception, but the firm still met problems characteristic of the industry. Management quarreled with shippers over the value of ore and bullion, and the imprecise assaying techniques of the day exacerbated the situation. Downs, Lowe, and other officials made periodic trips to mining camps to smooth ruffled feelings, but frustration, bitterness, and the lure of better returns sometimes allowed rivals to take business away. The enterprise occasionally had trouble paying mineowners because bankers in Omaha and their counterparts farther west were wary of each other's checks. And from time to time winter snows blocked ore shipments and compelled the firm to blow out furnaces and lay off workers.[31]

During the seventies, when many smelting companies filed bankruptcy petitions, the Omaha firm expanded its activities. Downs and his successors as president, Charles B. Rustin and Charles W. Meade, integrated operations by installing more furnaces, building a refinery, and establishing a far-flung network of ore buyers to ensure a continuous supply of mineral. The credit rating of the enterprise remained impeccable, and the firm even lent money to mining companies to increase their production. The smelter tapped many ore markets, but its nearest source lay in Colorado. When the minerals industry there changed radically in the late 1870s, the men of Omaha were well positioned to take advantage of new opportunities.[32]

Less than a year after the Omaha Company went into business, a rival appeared in St. Louis. The leading spirit in the venture was Edwin Harrison, an intense man whose six foot seven-inch frame commanded attention wherever he went. Harrison's background had prepared him well for a career in the minerals industry. The son of a prosperous businessman, he had attended schools in the United States and Europe before enrolling in the Lawrence Scientific School of Harvard University, where he studied under the renowned naturalist Louis Agassiz, who regarded Harrison as one of his ablest pupils. Graduating in 1856, Harrison found employment with the Missouri Geological Survey, which provided him with practical experience to complement his theoretical education.[33]

Harrison remained with the survey only a short time, for he was inclined toward a career in business. The year 1860 found him in Santa Fe as an agent for one of his father's trading ventures. He worked there for two years before returning to St. Louis to enter the iron industry with his father and several former fur traders who had launched the Harrison, Chouteau, and Valle Iron Company, which owned the Laclede Rolling

Mills. After his father died in 1870, the directors elected Harrison vice-president and later president. About the same time he was chosen president of the Iron Mountain Company, which was removing an entire hill near St. Louis to provide ore for the Laclede furnaces. With his scientific background, experience in heavy industry, and ready access to venture capital, Harrison held a unique position for launching a silver-lead smelting firm should railroads ever connect St. Louis to western mining camps. This link came in 1870 when the Kansas Pacific line ran its first train into Denver.[34]

On May 10, 1871, Harrison brought together a group of local businessmen who organized the St. Louis Smelting and Refining Company. They set the capitalization at $250,000, divided into 2,500 shares, and sold about one-third of the stock. As the chief entrepreneur in the venture, Harrison was elected president. The vice-presidency went to George W. Chadbourne, head of the St. Louis Shot Tower Company, which used large quantities of lead. The remaining officers and directors included Jonathan B. Maude, Martin Collins, J. L. Howard, and two bankers, George H. Loker and William L. Loker.[35]

Like all successful smelting firms, the St. Louis enterprise found experienced engineers to build and operate its plant. As his metallurgist Harrison hired Augustus Steitz, an American who had studied at Freiberg, examined mines in Colorado, and smelted ores with a blast furnace at Argenta, Montana. To construct a refinery, Harrison engaged August R. Meyer, another American-born graduate of the Bergakademie. His work for the enterprise launched one of the great careers in the industry, but his preeminence lay well in the future. Arthur H. Meyer, presumably no relation, became assayer.[36]

The staff recruited, Harrison and his colleagues purchased an industrial tract about five miles outside St. Louis and set to work. J. L. Howard, who became manager of the plant, directed construction while Steitz and the two Meyers supervised technical details. In July, Howard and Chadbourne left for Colorado and Utah to purchase ores. They found a ready supply, signed contracts with mining companies, and made arrangements with railroads to ship the product to St. Louis. Once the smelter acquired adequate supplies of silver-bearing galena, iron flux, and coking coal, Steitz blew in the Piltz furnaces.[37]

Supported by the "soundest and most cautious capital," the St. Louis company proved an immediate success. Railroads radiating from St. Louis gave the firm access to fuel, ores, and flux. The old lead-producing districts of Missouri provided high-grade galena whenever the lead content in western minerals fell below the optimum requirements for reduction. Chadbourne's Shot Tower Company and other local manufacturers made it easy for Harrison to market lead. With sampling agencies in western mining camps and a smelter and refinery in St. Louis, this well-integrated firm earned large profits and paid substantial dividends. Like their counterparts in Omaha, the St. Louis smeltermen were ideally situated to take advantage of new opportunities in Colorado when the minerals industry altered course there in the late 1870s.[38]

In the years after their founding, the Omaha and St. Louis companies competed for the relatively small production of silver-lead ores mined in the Rocky Mountains. The struggle centered largely in the valley of South Clear Creek, but competition ensued on a lesser scale in South Park and in the upper Arkansas Valley. While these two firms and a few rivals in Colorado, in the East, and in Europe smelted the territory's production of galena, mineowners still demanded the construction of local plants because most of their output consisted of lower grades of ore unable to bear the cost of transportation down from the mountains and across the plains. And by the late 1870s they were about to get their wish.[39]

Chapter 4

"The Realms of Puff"

IN 1858 SOME PROSPECTORS HEADING FOR THE NEW EL DORADO decided to eschew the well-publicized gravels of Clear Creek in favor of the unknown country in the Arkansas Valley. Some parties ascended the river from Bent's Fort. Others crossed the Mosquito Range from South Park. Still more arrived by way of Fremont Pass or Tennessee Pass. Regardless of the route they traveled, this was an arduous passage, for the headwaters of the Arkansas rose amid the highest peaks of the Rocky Mountains. Once in the river valley, these fifty-niners fell to work along the floodplain. Their supplies and fortitude dwindled as summer passed, but they still unearthed traces of placer gold at Kelly's Bar, Cache Creek, Georgia Bar, and other nondescript sites now long forgotten.

When the first snows of winter blew down the valley, the prospectors retreated from the mountains to spend their winnings and replenish their outfits for another season. Many went to "Denver City," still hardly more than a few tents, log cabins, and tepees huddled along the banks of the South Platte River. Like the Cherry Creek placers that touched off the gold rush, none of the discoveries on the Arkansas really amounted to much, but stories of easy wealth to be found in the valley abounded in Denver that winter. During the long months of inactivity, as the miners shivered on the cold, windy plain, flakes of placer gold grew by inference, suggestion, and fantasy into the harbingers of a real bonanza. And so early in 1860 new companies of fortune hunters, disappointed by their luck at the Gregory diggings, set out for the Arkansas country even before the snow was off the trails.

One party of prospectors included a weathered California miner named Abe Lee. They headed for the site of discoveries made the year before, but after a little work at each they knew the older placers held little promise. So they continued to ascend the river. Sometimes they traveled along the east bank, sometimes the west. They passed Twin Lakes, then Mount Elbert and Mount Massive, Colorado's two highest peaks, the snowcapped sentinels that guarded the riches about to be uncovered—and disregarded. It was now April. Months of hard work and frustration lay behind Lee's party when they finally came to a narrow gulch filled with brush, mesquite, and fir trees. They dug through five-foot snowdrifts to a partially frozen streambed, and soon after they had begun panning in the frigid water, Lee spied a few flakes of gold. He was exultant and—so the story goes—exclaimed that this placer was as rich as the placers of California. He was wrong, but no matter. His remark gave the district its name.[1]

Pay from the diggings in California Gulch surpassed the return from sites farther down the river, the news traveled fast, and the boom was on. Before long the newcomers had staked out claims for five miles up and down the ravine, created a mining district, and laid out the rude hamlets of Oro City and Sacramento City. Some say that ten thousand newcomers arrived in the next year, but this figure seems vastly exaggerated like most associated with mining booms. A better guess would be one thousand. One was Horace Austin Warner Tabor, a strong-willed fifty-niner who came with his wife Augusta and several friends in hope of finding more gold than he had found the year before at the Gregory placers.[2]

Despite the expectations of so many prospectors, the dream

of finding sudden wealth in California Gulch proved to be just another will-o'-the-wisp. There were not enough worthwhile claims to accommodate the hundreds who arrived in search of gold. And there were other problems as well. The mountain runoff could not supply enough clear water for the scores of pans, cradles, and sluices working the placers. By the time the stream ran down to the lower claims, the water had become thick, brackish, and wholly unfit for placering—"liquid mud" the miners called it. Then there was another source of vexation—the large pieces of iron-stained rock that had to be picked out of the sluice boxes and thrown away by hand. And, finally, throughout the Gulch lay another substance politely known as "black cement" or, more profanely, as "the damned blue stuff" that clogged sluices and frustrated the search for gold. These troubles proved too much for many miners, and in the fall of 1860 the less determined packed up and left, never to return. A few more resolute stayed on to confront the harsh winter and work their claims as weather permitted.[3]

Another full season for prospecting opened in 1861, but the backbreaking toil of placering for little or no "pay" dampened enthusiasm and throttled hopes. Population and production fell. In August two miners located a deposit of lead carbonate ore that held substantial amounts of silver, according to assays later made in Chicago, but this report, if not apocryphal, was quickly forgotten. Even so, the prospectors had no way to process "the damned blue stuff," and no one perceived the vast extent of the riches lying in the hills around Oro City. As 1861 drew to a wintry close, more miners drifted away. Even the Tabors moved on, crossing the Mosquito Range to Buckskin Joe, a more prosperous camp on the rim of South Park.[4]

For the next fifteen years California Gulch sank into the backwash of life in Colorado as the search for wealth focused on Central City, South Clear Creek, and South Park. A few hardy miners remained on the upper Arkansas scraping together a sparse living with pans and sluices. Some turned to hydraulic mining in hope of increasing their return, if not finding the big bonanza, but the feverish boom days of 1860 remained little more than a pleasant memory that grew ever more distant as the years passed. Even so indefatigable an investigator as Nathaniel P. Hill passed up a chance to visit the district when he made his way past Twin Lakes to Red Mountain in 1864.[5]

The desultory prospecting that continued during those languid times finally prompted some change. In 1868, a full eight years after Abe Lee had named the district, Charles J. Mullen and Cooper Smith found the first gold lode high on a ridge in California Gulch. This claim, named the Printer Boy, passed into the control of the Boston and Philadelphia Mining Company, which placed the property in the hands of J. Marshall Paul, a mining engineer of questionable reputation and less ability. He opened the mine and erected a stamp mill at Oro City, but his leadership proved indifferent, and the firm later replaced him with William H. James, a far abler manager known for his work on South Clear Creek. Nonetheless, the $600,000 to $800,000 from the Printer Boy over the next decade provided the district with much of its meager prosperity.[6]

The discovery of the Printer Boy brought new enthusiasm to mining on the North Arkansas. Prospectors staked many a claim on the timbered hills, rocky gulches, and steep ravines, but most of the claims held little gold. The primary mineral was a silver-bearing lead carbonate, "the damned blue stuff" for which there was little market. A. Y. Corman and several friends uncovered a number of deposits, one of which he named the A. Y. for himself, another the Minnie, for his wife. A Captain S. D. Breece opened the Berry mine on what came to be known as Breece Hill, a short distance south of California Gulch. And across the valley on Tennessee Pass a group of entrepreneurs developed the Homestake mine. The location of the Printer Boy also induced the Tabors to return to Oro City, where they opened a store managed largely by Augusta, while Horace engaged in grubstaking, speculating, and mining, though with little success.[7]

What plagued development more than anything else was the lack of an ore market. The Homestake mine shipped some of its production to a new smelter at Golden, but the large amount of nickel in the rock hindered the operation of both firms. Other companies sent small consignments of mineral to eastern smelters, but the astronomical freight costs created by the extreme isolation of California Gulch retarded development. In 1873 Tabor and other miners tried to raise local capital to erect a smelter, but the effort came to nothing. Three years later, in his last annual report on western mining, Rossiter Raymond noted that the lack of an ore market still hindered all mining operations in the district.[8]

Yet, even as Raymond composed his thoughts, change was coming to the upper Arkansas. At long last the demand for a local smelter induced a group of entrepreneurs from Ohio to found the Malta Smelting and Mining Company, a firm that intended to build works at the new village of Malta, a few miles from Oro City. The enterprise was incorporated in March 1875 with a nominal capitalization of $175,000, though there is no evidence that the founders ever raised this sum. Nevertheless, they hired Emile Loescher, a German-born metallurgist, to examine the terrain near Malta and erect a plant. He selected a site on a hillside so that ores could flow downhill in the course of reduction, put up a building, and installed a reverbertory furnace for roasting and a blast furnace for smelting. Raymond was impressed by the plant and hoped it would not be crippled by lack of ores.[9]

Loescher commenced operations in 1876. He drew silver-lead ores from mines in California Gulch and from the Homestake property across the valley, but their production could not sustain the works because of inadequate development during the long time when there was no local market for mineral. Loescher also found himself hampered by a scarcity of coke and by low-quality firebrick. He and his financial backers persisted throughout the summer and into 1877, but the smelter's operations remained sporadic.[10]

Meanwhile, the iron-stained rock that had long fouled the miners' sluice boxes aroused the curiosity of two veteran prospectors from South Park. In April 1874 Alvinus B. Wood crossed the Mosquito Range to work the Starr placer in California Gulch. Wood was familiar with both silver-lead and pyritic ores, and the peculiar mineral near Oro City sparked his interest.

He took samples back across the range to Alma, where they were assayed by Hermann Beeger of the Boston and Colorado smelter. Beeger found that besides iron the rock held 27 percent lead and fifteen ounces of silver per ton—too much lead and too little silver for the Alma furnaces.[11]

Undaunted, Wood formed a partnership with William H. Stevens, Hill's disenchanted partner in the Alma Pool Association. The two men then returned to Oro City to look for the iron-stained outcropping of ore bodies in the wooded hills above town. Over the next year they located what became the Lime, Rock, and Dome mines—discoveries that eventually led them to the far richer Iron Silver mine on what became known as Iron Hill. Wood and Stevens knew that the original placer claims, though long abandoned, were valid for another year. Shrewd and experienced in a cutthroat business, they waited in silence. Then in the spring of 1876 they quietly acquired the properties that one day made up the Iron Silver Mining Company, one of Colorado's great producers. Once in control, Wood and Stevens hired a group of miners to sink shafts into the ore bodies, but for as long as possible they kept their employees in the dark about the true value of the deposits.[12]

Wood and Stevens knew they had a valuable group of mining properties—but only if they had an ore market. In the summer of 1876 they returned to Alma, where August R. Meyer, a slender man with a thick moustache, had just established a purchasing agency for the St. Louis Smelting and Refining Company. Using their rich samples as a lure, Wood and Stevens persuaded him to travel from his headquarters over the mountains to Oro City. Once he was on the North Arkansas, the potential for profit impressed Meyer, and he purchased from two to three hundred tons of mineral drawn largely from the Rock mine. Local transportation was wholly inadequate for getting the ore out of the mountains, but Meyer was resourceful. He went to New Mexico, obtained several teams of oxen, and had the mineral hauled to the railhead at Colorado Springs before the winter snows arrived. The cost of shipping the ore to St. Louis was high, but the smelting returns gave a profit to all involved. The news gave tremendous impetus to mining in California Gulch. By the end of 1876 Colorado was on the eve of a great new bonanza.[13]

Early knowledge of such wealth gave Edwin Harrison and his associates an exceptional opportunity. Early in 1877 they shifted Meyer's base of operations over the mountains from Alma to Oro City, where they set him up in business as the A. R. Meyer & Company Ore Milling & Sampling Company, a name that must have pleased Meyer, who was inclined to be garrulous. But this handsome, moustachioed man with his debonair dress was also a skillful businessman with a Calvinist devotion to work—just the person Harrison wanted on the North Arkansas. It was not long before the sampling agency was shipping large quantities of mineral to the furnaces in St. Louis, but only rich ores could bear the transportation costs. From his post Meyer sent compelling reports to St. Louis, for his thorough letters left nothing to conjecture—development work on the mines appeared promising, demand for a smelter was high, and the Malta works were unequal to the task.

Meyer's reports prompted Harrison and his associates to talk of expansion—a plant in California Gulch. They knew that it would be all but impossible to get adequate supplies of coke because the nearest railroad lay eighty miles away across the mountains—witness the problems of the Malta Company—but they thought local forests might provide enough charcoal. The mines of Wood and Stevens held vast quantities of flux, whose silver meant the ore would not have to be purchased at a loss. And since Wood and Stevens were urging Harrison to build a smelter, they might accept a lower return than otherwise, at least in the beginning. Limestone was also available. The St. Louis enterprise was also in sound financial condition, its credit excellent, its staff skilled. Having studied the situation as best they could, Harrison and his cohorts decided to build a smelter near Oro City.[14]

In June, Harrison arrived in Denver to let contracts and confer with Meyer, now general manager of operations in Colorado. As soon as building materials arrived from St. Louis and Chicago, the two men left for the North Arkansas to begin construction on a two-and-one-half-acre site below the mouth of California Gulch, a few miles from Oro City. To acquire more land, Meyer located a placer that he sold to Thomas Starr, a longtime resident of the district. Starr combined the claim with a tract of his own, secured a mineral patent on 165 acres of land, and sold the estate back to the St. Louis Company. These transactions later prompted a series of lawsuits, all won by the company, much to its advantage, for a new community was to grow up on the land acquired. But in 1877 legal battles lay well in the future. In October, as fall gave way to winter ten thousand feet above sea level, Meyer completed a smelter that looked like a large barn with a short smokestack but was fitted with coke and ore bins, an assaying room, storehouse, roasting unit, and one large blast furnace with a capacity of forty tons daily. Later that month Meyer began operations at the new Harrison Reduction Works.[15]

It was Meyer's job to coordinate the work of the sampling agency, the smelter, and the plant in St. Louis. Under a system devised shortly before the furnace was set in blast, Meyer had his agency assay all ore samples shipped by mineowners. If the two parties agreed on the content of silver and lead as well as the contract offered by the smelting company, he purchased the consignment on behalf of the St. Louis firm. Because the Harrison works had a relatively small capacity, he shipped the high-grade ores directly to St. Louis. Minerals of lesser value, unable to bear the transportation charges, went to the furnaces on the North Arkansas. In the first months of operation the Harrison works reduced every three tons of silver-bearing carbonates to one ton of lead-base bullion, attesting to the richness of the surface ores mined from the Iron Silver, Argentine, and other properties. Meyer employed sixteen teams of oxen to haul ore and bullion over the mountains to Colorado Springs and return with El Moro coke.[16]

As the first stirrings of a boom marshaled strength in 1877, the only towns in the mining region were Malta and Oro City, but they achieved little growth. Most new construction took place a few miles away around the Harrison Works at the foot of Carbonate Hill. For a time the new hamlet had no name, and mail came addressed to Oro City or Malta. But on January 10, 1878, Meyer, Wood, Tabor, and others gathered to organize a town. As the story goes, Meyer proposed the names Cerussite and Agassiz, but his colleagues rejected both as too erudite for a mining camp. Someone suggested Harrison, but the conferees dismissed this, too, although the town later named its main thoroughfare after the president of the smelting company. Finally, Wood advanced the name Lead City. This brought a more favorable response until someone noted that the hamlet might be confused with another mining camp in the Black Hills of Dakota Territory. The group then changed the name to Leadville. Tabor was promptly elected the first mayor, although Meyer laid out much of the town—or so he later claimed.[17]

The rich mines on Iron and Carbonate hills won prominence for Leadville, but it was the discovery of mineral on Fryer Hill and the charismatic personality of Horace Tabor that gave the town its fame. Late in 1877, as older properties began to draw thousands of fortune hunters to the upper Arkansas, August Rische and George Hook asked Tabor for a grubstake. Having helped many prospectors over the years, he gave them seventeen dollars worth of equipment in return for one-third of whatever they found. Rische and Hook set out for the hill where George Fryer had recently found the Discovery claim. They staked out a property nearby and began digging. Why they happened to pick this particular spot remains conjectural and forms an integral part of the Tabor legend. Nonetheless, at a depth of thirty feet the two prospectors struck the main ore body of what Hook named the Little Pittsburg mine in honor of his hometown. By great luck they had sunk a shaft at the only point on the hill where the rich silver-bearing mineral came near the surface.

Tabor's star was ascendant. He and his partners immediately commenced development on what proved to be a bonanza claim. Then, in a series of complicated transactions Tabor and two Denver bankers, Jerome B. Chaffee and David H. Moffat, bought out Rische and Hook, acquired several adjacent properties for protection, and incorporated the Little

August R. Meyer in 1871, as a student at Freiberg, Saxony, shortly before he went to work for the St. Louis Smelting and Refining Company. Colorado Historical Society, Denver.

Pittsburg Consolidated Mining Company, the basis of the storekeeper's fortune.[18]

One bonanza might have satisfied most individuals, but Fortune smiled twice upon Tabor in 1878. A few weeks after Rische and Hook found the Little Pittsburg, Tabor's suppliers in Denver asked him to acquire a promising claim for them.

Tabor in turn commissioned William "Chicken Bill" Lovell, of nefarious reputation, to prospect in another section of Fryer Hill. According to the story, "Chicken Bill" sank an exploratory shaft on an old claim known as the Chrysolite and then "salted" his work with ore either borrowed or stolen from the Little Pittsburg. Completely fooled, Tabor purchased the claim for $900 and set out for Denver to resell the property for $40,000. "Chicken Bill," however, bragged too much about his chicanery, the news preceded Tabor to Denver, and his wholesalers refused to buy. Undaunted, though perhaps embarrassed, Tabor hired a crew to dig deeper in hope of finding mineral. His luck still held. The miners had gone only a short distance when they struck another rich ore body. Tabor and several partners then acquired several nearby claims and incorporated the group as the Chrysolite Silver Mining Company, a large producer of metal in the next few years and another milestone in the life and legend of Horace Tabor.[19]

The growth of Leadville was phenomenal even by the exaggerated standards of mining camps. When Harrison and Meyer erected their first blast furnace in 1877, the village had just two hundred inhabitants, and a correspondent for the *Engineering and Mining Journal* casually mentioned that the district was "laying the groundwork for more extensive operations in the future." But a year later in the midst of boom another writer for the *Journal* found that the most difficult part of his task would be "to keep within the bounds of reason . . . and not say so much . . . as to border on the realms of puff." The *Mining and Scientific Press* talked about the infant city "with its closely built streets, its bustle of trade, its throng of teams that fill and block the way, and its surging masses of humanity that move in ceaseless currents." This was Leadville, where life was intensified, where "everybody who [had] anything to do was on the jump," where "the rasp of the saw and tattoo of the hammer [could be] heard from daylight to dark seven days in the week." Miners, tradesmen, people of all sorts had to elbow and push their way through the crowds on the sidewalks or wait for an opening in the teams and vehicles in order to cross the street. This was a town "overrun with

Leadville. The Harrison Reduction Works stand at the left, August R. Meyer & Company on the right. Gurnsey, photographer. Colorado Historical Society, Denver.

multitudes of half famished, diseased, and nearly desperate people."[20]

By 1880 the population had soared to nearly fifteen thousand, making Leadville the second largest city in Colorado. What in 1877 had been a tiny, obscure mining camp with no name had emerged as a mountain metropolis with twenty-eight miles of streets, gaslights, waterworks, several banks, three hospitals, thirteen schools, 114 saloons, uncounted bordellos, and fifteen smelters with thirty-seven blast furnaces. The surge coincided with the decline of Eureka, Nevada, as a lead producer and Virginia City, Nevada, as a silver producer. By 1880 Leadville was supreme as the source of silver and lead in the United States.[21]

The large amount of venture capital that flowed in to open the Robert E. Lee, Morning Star, Evening Star, A. Y., Minnie, Highland Chief, Oro La Plata, and countless other mines produced a startling result on the tally sheet. In 1877 Lake County yielded metal worth a scant $670,000, a poor fourth behind Gilpin, Clear Creek, and Park counties in Colorado's grand output of $6,900,000. But the next year Leadville's production surged to nearly $2,500,000, which surpassed all other counties in the state's total of $9,200,000. And this was only a pittance compared with what Lake County was to produce in the years to come.[22]

The soaring production overwhelmed the Harrison and Malta works, but this made little difference, for agents like Meyer were ready to buy ores for eastern smelters. Late in 1877 Nathaniel Witherell and Theodore Berdell opened an office in Malta for the Omaha Smelting and Refining Company, which did not want to see its rival in St. Louis monopolize the trade. In spite of the heavy winter snows, Witherell and Berdell shipped large quantities of carbonates to the furnaces in Omaha. Then early in 1878 William H. James, the former superintendent of the Printer Boy, created a partnership with his old friend Edward Eddy, who had spent nearly twenty years in the mining industry of South Clear Creek. They opened the Leadville Sampling works in July, but, unlike Meyer, Berdell, and Witherell, who managed agencies that provided a measure of vertical integration to parent firms, Eddy and James operated an independent sampler that sold ores to many smelting companies. They sent some consignments to Omaha, some to St. Louis, and some to Pueblo, Colorado, where the firm of Mather & Geist had just built a plant to tap the Leadville market.[23]

Despite the work of such ore buyers, the surging output of the mines created a strong demand for local smelters. The major reason was that the primitive transportation system in the central Rockies could not ship the enormous tonnage of mineral to eastern works, although this was somewhat beside the point, since production soon exceeded the capacity of the entire American smelting industry. Only the Malta and Harrison works were in operation when the boom commenced in 1877, but two more enterprises entered the industry the next year, when the mines produced $2,500,000. The next year eleven more smelters came on stream as Leadville's output soared to $11,300,000. The growth of the mines and smelters—as well as the tepid symbiotic relationship between them—altered the structure of the minerals industry and the state's economy.[24]

The Malta enterprise that had erected the original smelter on the North Arkansas tried to share in the bonanza. Ever since Loescher had built the plant, operations had been sporadic, but in 1878 the firm purchased about five hundred tons of ore from the Adelaide mine and ran steadily for a time. In April, however, financial setbacks and poor construction ended the career of the Ohioans, who "retired from the field," in the parlance of one commentator, despite an investment estimated at $80,000.[25]

Yet the demand for smelters was too great for the works to sit idle for long. The plant passed into the hands of another group of entrepreneurs composed of J. B. Dixon, his brothers, their wives, and several in-laws. They leased the plant, created a partnership on September 1, 1878, and purchased the works for $10,000. What any of them knew about ore reduction is conjectural. Dixon himself had been in the theatrical business at Indianapolis, although he ran a foundry as a sideline. Despite this rather novel background, the new owners had the sagacity to hire Franz Fohr, one of the ablest metallurgists in the United States.[26]

Everyone who knew Fohr, it seems, remarked about his unusual personality—cold, austere, forbidding, and secretive about his knowledge of metallurgy. "Pity he is such a fool," wrote one unfriendly observer. Yet to those who got to know him, like Henry Wood, the ore buyer who used to ride out with Fohr to Twin Lakes on Sundays, he was much friendlier and more communicative. Fohr liked to say that his father had told him to divide his life into three periods—the first twenty-five years devoted to study (which he had done), the second twenty-five devoted to work (which he was doing), and the last twenty-five devoted to pleasure (which he would do). Like so many metallurgists, he was a Freiberg graduate but had worked for some time in the United States, including a short stint in the old smelter at Saints John.[27]

Under the new regime, operations at the Malta plant went

well for a time. Fohr put the furnaces back into blast, shipped substantial amounts of bullion to eastern refiners, and in 1879 erected two more smelting furnaces. In obtaining their large supplies of ore, Dixon and his partners signed a contract with Meyer, who agreed to purchase carbonate for the Malta works at the same rates he did for the St. Louis enterprise. In return, Dixon agreed to send his bullion back to Meyer for shipment to the refinery in St. Louis. This arrangement, however, proved unfortunate for Dixon and his associates. In October, Meyer secured an attachment against the firm for more than $9,000, a sign of continued financial problems that soon led to the company's downfall. Despite the mounting internal troubles, Dixon and Fohr produced about $984,000 in bullion, about 10 percent of Leadville's output that year.[28]

Unlike the Malta Company, the St. Louis enterprise prospered during the boom. Meyer effectively coordinated operations, shipping high-grade ores to St. Louis, sending lower grades to the Harrison works, and farming out other consignments to new firms. As production grew, Harrison and his associates appropriated funds so that Meyer could build a second blast furnace, which he did in August 1878, doubling the capacity of the plant. Actual operations went smoothly under the direction of James Brierton, the first superintendent, and his assistant, George A. Hynes, although some rather ludicrous problems developed when George W. Chadbourne, one of the directors, sent his son to Leadville for a summer job. He read the incoming and outgoing correspondence, insulted the local managers, and made life unpleasant until he returned home. Then on January 1, 1879, Brierton left, and the enterprise promoted Hynes to superintendent. Yet he was unhappy in the job. He was bothered by the altitude and preferred the comfortable, cosmopolitan atmosphere of St. Louis to the rough-hewn ways of the carbonate camp. He also feared that Chadbourne's son might return for another summer. Nonetheless, Hynes did commendable work as Meyer's lieutenant. In the bonanza year of 1879 the Harrison works produced bullion worth $941,000 from ores averaging $116 a ton. This was an impressive yield, but the smelting industry had grown so rapidly that the output was only the fifth best in town.[29]

In addition to the business of ore reduction, Harrison and his colleagues developed some mining interests. In March 1878 they paid a reported $200,000 for the Camp Bird, Charlestown, and Pine lodes on Carbonate Hill and organized them into the Argentine Mining Company, which Meyer operated from his offices at the smelting works. Although the ores were rich in lead, the silver content fell off as Meyer pushed development underground. By good luck, however, the mineral held small amounts of gold, and for this reason the Harrison plant was the only smelter in camp to process it, albeit in tiny quantities. Arthur H. Meyer, who had succeeded Augustus Steitz as metallurgist in St. Louis, also formed a mining company in conjunction with Chadbourne and the Loker brothers. These enterprises gave the smelting firm a degree of informal vertical integration characteristic of the industry.

Harrison also engaged in other ventures. He was active in bringing the Denver South Park & Pacific Railway to Leadville, and he helped organize the Harrison Hook and Ladder Company to provide the town with some protection against the ever-present danger of fire. He also toyed with the idea of erecting a smelter at Alpine, a new camp about eighty miles southwest of Leadville, where he and Tabor speculated briefly in mining properties. Yet in spite of Harrison's success within and outside the reduction business, neither he nor Meyer was the most important smelterman in Leadville by the end of 1879. This honor belonged to James B. Grant, who had just launched his extraordinary career in the minerals industry.[30]

By origin Grant was a southerner, born in Russell County, Alabama, in 1849, the year of the California gold rush. He grew up in the Old South and served briefly with a Confederate regiment toward the close of the Civil War. With the southern economy in ruins, he set out for Davenport, Iowa, where he persuaded a well-to-do uncle to finance his education, first at what are now Iowa State and Cornell universities and then at the Bergakademie of Freiberg, Saxony. Grant spent three years in Europe pursuing the study of mining engineering and metallurgy. Then he embarked on a round-the-world cruise at least in part to examine the emerging minerals industries of Australia and New Zealand. Once back in the United States, he took up his profession as a mining engineer at Central City, Colorado. And it was here that he developed his passion for duck hunting and grew the long beard that counterbalanced his already balding pate.[31]

Grant acquired a measure of experience in the famous gold town, but it was the rise of the carbonate camp that launched his career as an industrial capitalist. In the winter of 1877 the Pueblo & Oro Railroad had him examine the mines at Leadville to determine if future production would justify building a line up the Arkansas Valley. After completing his survey as the snows blanketed the high country, Grant wrote his uncle that he had "never seen such extensive deposits of silver-bearing lead ores." A railroad would surely pay large dividends. The line published this letter in a promotional tract, but Grant's activities indicated that his note reflected his real beliefs. What was more, like Nathaniel P. Hill more than a decade before, he had used his journey to the North Arkansas for more than the purposes outlined by the Pueblo & Oro Railroad.[32]

In the spring of 1878 Grant returned to Iowa to convince his uncle that smelting ores in the central Rockies would be profitable. His education at Freiberg had given him the requisite technical knowledge, but he did not have the capital. His uncle did. James Grant had long been one of Davenport's leading citizens. Although he liked to pass himself off as a farmer, he was in fact a very successful lawyer, judge, and industrialist, a man who had served as president of the Chicago and Rock Island Railroad and who possessed a fortune worth perhaps as much as $500,000. He listened carefully to his nephew's arguments, then he agreed to form a partnership and finance the project, interested as he was in both his nephew's career and the chance for profit.[33]

Assured of adequate capital, Grant returned to Leadville. He purchased a tract of land on Front Street west of Leiter Avenue and broke ground for the smelter in June. Construction went slowly, however, because the primitive transportation system frustrated quick deliveries to Leadville. Railroads carried fire-brick, coke, machinery, and blast furnaces made by Fraser and Chalmers in Chicago to the end of the line, wherever that happened to be, but from there teamsters had to haul the material over the rough, mountainous terrain to "cloud city." This was time-consuming and expensive, and it forced Grant to delay the start of operations several times. Finally, on October 3 he set his first Piltz furnace in blast. By the end of the year this unit yielded about three hundred tons of lead-based bullion that Grant sent over the mountains and across the plains to Omaha for refining. These shipments marked the beginning of a close relationship between Grant's firm and the Omaha Company.[34]

Once under way, Grant's expansion was remarkable. Early in 1879 he erected three more Piltz furnaces, which made his plant the largest in Leadville after only a few months of operation. Then in July he signed a contract to purchase the entire production of Tabor's Little Pittsburg Company—twenty thousand tons of ore—from August 1, 1879, to January 1, 1880. This demanded further expansion. To fulfill his obligations to Tabor and other mineowners, Grant had to double his ore capacity from 80 to 160 tons daily. He telegraphed Chicago asking Fraser and Chalmers to ship four additional Piltz units, but the firm replied that it could send only two. Grant then signed a contract for two rectangular (Raschette) models with the Council Bluffs Iron Works of Iowa. The agreement specified that the furnaces had to arrive in Leadville by the middle of August, but the Council Bluffs enterprise failed to fulfill its obligations. Not until fall did Grant receive two defective units that cost him about $300 a day in revenues. He finally installed the two Fraser and Chalmers models in October, and he acquired two other furnaces by the end of the year.

Furnace problems notwithstanding, Grant's production of silver-lead bullion rose steadily throughout 1879. By the end of the year he had shipped $2,400,000 worth of metals, nearly one-fourth of Leadville's output and only $11,000 below the yield of the Boston and Colorado Company, the state's foremost reduction firm. With an investment of about $160,000, furnished mostly by his uncle, Grant had erected the largest smelter in Leadville.

James B. Grant, of course, had not done all this by himself. Early in the year Judge Grant had moved to Leadville to participate in entrepreneurial decisions and handle legal affairs, even though the rigors of the climate must have been hard on a man in his sixties. The firm also engaged Malvern W. Iles to serve as assayer. A fortuitous choice, Iles was a graduate of the Columbia School of Mines and proved himself an able employee. As the younger Grant devoted an ever larger portion of his time to business affairs, Iles assumed greater metallurgical responsibilities and emerged as one of the leading technical experts in the industry.[35]

As Leadville's rising production made headlines across the nation, Nathaniel Witherell and Theodore Berdell, who bought ores for the Omaha smelter, also considered entering the reduction industry. In May 1878, just before Grant broke ground for his works, they had their chemist, Henry B. Yelita, conduct experiments to determine if they could process the huge quantity of silver-lead ores that could not bear the cost of transportation to Nebraska. The results were favorable, and although Yelita left the company that fall, Witherell and Berdell pressed forward with the project in concert with their silent partner, Charles B. Rustin, a former president of the Omaha firm. The three men bought a parcel of land outside the city limits, erected a smelting plant, and set their initial furnace in blast only a few days after Grant himself began operations.

As firm believers in the great future of the carbonate camp, Witherell, Berdell, and Rustin expanded almost immediately. In January 1879 they installed a second smelting furnace. This unit also proved successful, and in June, as Leadville's production climbed to new heights, the partners added a third model, bringing the capacity of the works to about ninety tons daily. The firm shipped its bullion to the Balbach plant in New Jersey, where the product was refined into silver and lead.[36]

Yet Witherell, Berdell, and Rustin were still unsatisfied.

They saw even greater possibilities for profit if they integrated operations back into mining, but to do this they needed capital—far more capital than they had. Early in 1879 they opened negotiations with investors in New York, always a ready market for mining securities. Talks, correspondence, and legal maneuvering continued throughout the spring, then culminated on June 11 with the formation of the La Plata Mining and Smelting Company. The trustees elected Rustin president, Witherell vice-president with offices in New York, and Berdell treasurer with headquarters in Leadville. The firm had a nominal capital stock of $2,000,000, divided into 200,000 shares, each with a par value of ten dollars; but when the enterprise listed its securities on the New York Stock Exchange, shares fell to four dollars each, suggesting that more than half the capital was "water." Even so, English investors purchased a majority of the outstanding stock.[37]

The La Plata enterprise moved quickly to consolidate operations. It acquired the smelting works of Witherell, Berdell, and Rustin; purchased the Gnesen, Slipper, Montgomery, and Oro La Plata mines; and installed a fourth blast furnace, increasing the reduction capacity to about 120 tons daily—second only to Grant's plant. Even though the company had its own supply of ores, it still purchased mineral from other shippers so as to prepare optimum furnace charges and to profit from the output of other rich properties. The year 1879 proved very successful. By the end of December the works had smelted 19,600 tons of ores into bullion worth nearly $1,700,000, nearly 15 percent of Leadville's yield. And during its first six months in business the La Plata firm declared two dividends, each of $45,000.[38]

As the news of the Leadville boom swirled across the West like a tornado, the camp stirred the imagination of the two most prominent smeltermen in Utah. Gustav Billing and Anton Eilers were part-owners of the Germania Smelting and Refining Company, whose plant at Flach's Station in the Salt Lake Valley was the largest in the territory. Billing had been associated with the enterprise since the early 1870s, when he arrived from Germany to direct operations. He had done fine work and acquired an excellent reputation in American financial circles. Yet, when technical problems beset the smelter, the real key to success had been Eilers's skill as a metallurgist.[39]

Unlike Meyer and Grant, who were virtually unknown before the Leadville boom, Eilers had built an outstanding reputation in mining circles. Born in the German state of Nassau in 1839, he grew up there and entered the University of Göttingen, where he remained for a time before going on to the mining academy of Clausthal. Sometime after completing his education in Europe, he emigrated to the United States, at least in part because of a family dispute. He had no job when he arrived, but he soon found employment as a mining engineer for the firm of Justus Adelberg and Rossiter W. Raymond, the beginning of a lifelong friendship with Raymond. Eilers spent the next few years traveling from the islands of the Caribbean to the mountains and deserts of the West. After three years of this arduous, if not dangerous work, he settled into a more sedentary life managing a copper mine and smelter in Virginia; but when Raymond became United States commissioner of mining statistics in 1869, he hired Eilers as deputy commissioner. Once again Eilers toured the mining districts of the West to gather much of the information that appeared in Raymond's massive annual reports.

Eilers took advantage of his travels to study the early American practice of silver-lead smelting—which he found none too good. When the American Institute of Mining Engineers began to publish its *Transactions*, Eilers wrote articles urging the industry to use more scientific methods, adopt better accounting techniques, improve the control of operations, and reduce the losses of metal. Yet his most important contribution was his development of new slag types that increased efficiency and reduced losses of metal. By the time the Germania Company hired him in 1876, Eilers had developed an exceptional reputation as a metallurgist, and his success in solving the technological problems of that firm increased his renown.

From behind his long dark beard, Eilers projected a shy, austere visage to the public; but in private he was warm, humorous, and intellectually stimulating. Thoroughness— *Gründlichkeit*, as he said—was his trademark, and this he expounded to his colleagues and demonstrated in his writings and in the personal risks he took in gathering data for Rossiter Raymond.[40]

Throughout 1878 Billing and Eilers discussed the possibility of erecting works in Leadville, assuming of course that the carbonate camp held genuine potential. Late in the year, as winter settled over the high country, Eilers set out to determine the longevity of the mines (as nearly as this could be done) and the feasibility of building a smelter in such an isolated region. Like Harrison, Grant, and others, he was quickly convinced the ore deposits held great promise. Once he returned to Salt Lake City, he conveyed his assessment to Billing, and the two men agreed to form a partnership and relocate on the North Arkansas.

The next year proved tumultuous. To obtain capital for the smelter, Billing and Eilers sold their interest in the Germania Company and apparently obtained loans from either New York or German bankers. During the winter they ordered machinery, furnaces, and other equipment so that it would be ready for shipment to Leadville as soon as the roads cleared. Haste was essential, for Harrison and Grant already held a lead, and still others were known to have plans for entering the industry. Billing and Eilers broke ground on land they had acquired in the valley south of Leadville, pushed construction in the spring, and set their initial furnace in blast on May 14. Operations were successful from the outset, and the firm soon built a second unit. By the end of 1879 the Utah smelter, as it was often called, had reduced nearly eleven thousand tons of ore into $1,080,000 worth of silver-lead bullion. This was the third largest production in camp and represented almost 10 percent of Leadville's output that year.[41]

As 1879 drew to a close, the Grant, La Plata, Utah, Malta, and Harrison works dominated the Leadville industry. Their aggregate production of $6,900,000 in silver-lead bullion represented two-thirds of the grand output of $11,300,000. Except for the Malta plant, each enterprise had experienced management and adequate sources of working capital. Of the fifteen smelters in town, these five were in the most advantageous position, but not all would survive the competitive struggle. Of the ten firms that formed a distinct second tier within the industry, three were particularly important because they later emerged as major processors. This, of course, was not obvious, for in 1879 the American Mining and Smelting Company, the firm of Cummings and Finn, and the Elgin Mining and Smelting Company produced very small quantities of bullion. In fact, their aggregate production fell short of the yield of the Harrison works, the smallest of the big five.[42]

The American Mining and Smelting Company came into being in March 1879 when it received a corporate charter under the laws of the state of Illinois. The first president was Caleb B. Wick, a Chicago businessman, but he was quickly succeeded by Henry I. Higgins, the original secretary and treasurer, who had done well as an iron and steel commission merchant in Chicago. The stockholders included some prominent railroad executives, notably Charles E. Perkins, vice-president of the Chicago, Burlington, and Quincy Railroad, which served Denver, and William B. Strong, vice-president of the Atchison, Topeka, and Santa Fe line, which ran trains to Pueblo. While the potential revenues from mining districts loomed large in the plans of railroad builders, mines and smelters offered ancillary opportunities for investment. In this case both the Burlington and the Santa Fe firms, as well as their executives, hoped to profit from the burgeoning minerals industry in central Colorado.[43]

Unlike many who entered the smelting industry, Higgins and Wick knew it was essential to hire an experienced metallurgist, particularly because they themselves knew so little about technical matters. After a short search they retained Otto H. Hahn, whose long, varied career had made him almost as well known as his friend Anton Eilers. Born in Thuringia in 1845, Hahn had studied mining engineering and metallurgy at the mining academy of Clausthal. Like Eilers, he emigrated to the United States in 1863 and found employment with Adelberg and Raymond. After serving on their staff for several years, he moved west and happened to be in Nevada when the opening of the silver-lead deposits at Eureka gave him a unique opportunity because he possessed the technical knowledge for smelting silver-lead ores. With Albert Arents, Winfield Scott Keyes, and others, he introduced the first economically remunerative use of blast furnaces in the West. Later he developed a vigorous metallurgical consulting business, wrote articles for the American Institute of Mining Engineers, and eventually worked for the Germania firm. His skills were in demand, and the American Mining and Smelting Company was lucky to secure them.[44]

Drawing upon their limited capital of $70,000, Higgins, Wick, and Hahn wasted little time in putting a smelter into operation. After rushing construction in the spring of 1879, Hahn set the initial furnace in blast on June 5. This proved successful, and in November the firm installed a second unit. Although the output that year was a modest $256,000 in bullion, the enterprise had a capable management that would prove itself able to compete with larger rivals. Its heyday would come later in the 1880s.[45]

A second Illinois corporation formed in 1879 was the Elgin Mining and Smelting Company, created by a group of entrepreneurs from the town of Elgin, outside Chicago. The president of the firm was Albert Sherwin, a major figure on the Leadville scene for the next two decades. S. D. Wilder served as secretary of the corporation, while F. C. Garbutt, an experienced metallurgist, held a seat on the board of directors. The firm had a nominal capitalization of $500,000, although it is doubtful that the officers ever raised anything approaching this figure. Nonetheless, Sherwin and his associates drew on what money they had to purchase land in Big Evans Gulch near Leadville and erect what many contemporary observers regarded as the model plant in camp. The enterprise blew in its first blast furnace on June 24, 1879, and produced some $387,000 worth of bullion by the end of the year. Though small, this production marked the start of a plant that would outlast nearly all others.[46]

Soon after the Elgin firm commenced operations, M. J.

Cummings and Nicholas Finn completed their works. Under the direction of Thomas McFarlane, the former superintendent of the Wyandotte smelter in Michigan, the firm set two furnaces in blast on July 25, 1879. Shortages of coke and water hampered operations for the rest of the year and limited production to $285,000, but Cummings and Finn had such confidence in the future that they installed a third smelting unit in December 1879 and a fourth in January 1880. Like the Elgin and American companies, the partnership enjoyed good management and had a competent metallurgist directing all technical procedures. This smelter, too, was well equipped for the competitive struggle in the years ahead.[47]

In addition to Billing and Eilers, Cummings and Finn, and the Elgin and American companies, seven other firms constructed smelting plants in 1879. They were the California Smelting Company; the partnership of Raymond, Sherman, and MacKay; the Leadville or Lizzie Smelter; the Ohio & Missouri Smelting Company; Gage, Hagaman & Company; and the Adelaide and Little Chief mining companies. But faulty construction, poor metallurgy, bad management, inadequate capital, and bad luck made their day in the industry a brief one.[48]

As these fifteen smelters pursued their individual destinies, each one found that fuel was an essential resource. Everyone knew that the best substance was El Moro coke, a product manufactured from coal mined near Trinidad in southern Colorado, about 225 miles away. A nearer, albeit inferior, fuel came from the town of Como, which lay on the main line of the Denver South Park & Pacific Railroad, then building toward Leadville. Yet in these early days metallurgical coke was a scarce, expensive commodity because of poor transportation and inadequate development within the coal industry. Prices in the carbonate camp ranged from twenty-five dollars a ton during the summer when trails were passable to as much as sixty dollars a ton in other seasons when rain, snow, and high water disrupted shipments across the mountains. Sometimes there was no coke to be had at any price.[49]

With coke at a premium, smelters turned to charcoal, and

woodcutters destroyed entire forests to meet the demand. Axmen and sawyers used flumes and troughs to send logs down the denuded hillsides. And although the smelters passed it by, Malta emerged as an important charcoal center. Its beehive kilns and white, conical ovens stood out in sharp relief against the blackened landscape that had been green with firs only a few years before.[50]

The dearth of coke forced the smelters to mix what they had with charcoal. This practice became universal in camp, but the thirty-seven blast furnaces proved so voracious that shippers could not supply adequate quantities of either fuel. Competition was intense. Operations at the Little Chief, Leadville, and other plants always remained sporadic because of shortages. And during the harsh winter of 1879 heavy snows created a freight blockade that produced what many called a coke famine. The shortage had a devastating effect on the marginal firms, but even the largest and most successful enterprises— Billing and Eilers, the Harrison Works, and Cummings and Finn—had to take smelting units out of blast. The only enterprise untouched by the coke famine was J. B. Grant & Company. During the previous summer Grant had used his own mule trains to haul enormous quantities of charcoal and coke to storage bins at the plant. This stockpile carried the smelter through the winter months.[51]

At the beginning of 1879 smelting charges were high because a buyer's market prevailed. The four plants in camp had a small capacity in comparison to the production of the mining companies. Under these conditions the smelters charged a basic reduction fee of thirty dollars a ton—before adjustments. Then they paid for 90 percent of the silver content whose specific value was determined by the daily quotation in New York City—about $1.08 an ounce. The processors paid fifty cents for each unit of lead, but only if the ore exceeded 30 percent; for each unit below this minimum they deducted fifty cents a ton. Price schedules like these gave the smelting firms about half the value of most ores, which in 1879 averaged about ninety-five dollars a ton.

By the end of the year, however, the price structure had

changed because the surge in smelter construction increased the number of furnaces in camp from eight to thirty-seven. Competition drove the fixed reduction fee down from thirty to twenty-five dollars a ton. At the same time the smelters increased their payment for silver to 95 percent of the New York quotation and offered twenty cents for each unit of lead regardless of its percentage in the mineral shipped. The new schedules now gave the mining companies about two-thirds of an ore's value, the reduction firms one-third.[52]

Fluxing ores had their own price schedule because they were so crucial in the reduction process. The opening of the Lime, Rock, and Dome mines by Wood and Stevens had helped persuade Edwin Harrison and his colleagues to erect a plant at the foot of California Gulch. The growth of the smelting industry created a huge demand for fluxing material, particularly ores from the Iron Silver mine, which held silver and thus lowered costs. Competition for such mineral was so intense that prices rose sharply. Smelters often purchased the mineral at the cost of reduction or even at a loss, a situation that benefited the shipper. Firms like the Iron Silver Company earned large profits and paid substantial dividends to their stockholders.[53]

In purchasing ores, smelting enterprises submitted bids on lots offered for sale. George Hubbard Holt, the manager of the Little Chief Mining Company, sold consignments to the Berdell and Witherell, Grant, and American firms while awaiting the completion of his own works. He readily switched companies according to the best offer he received or according to his opinion of the smelter's integrity—he frankly questioned the practices of Berdell and Witherell. Conversely, Billing and Eilers bought ores from the Evening Star, Amie, Chrysolite, Highland Chief, and other mines. Long-term contracts such as Grant's agreement with the Little Pittsburg Company were rare. Once many smelters had entered the industry, the bidding system worked to the advantage of mineowners. Smeltermen tried to collaborate on price schedules and form pools to maintain uniform rates, but such informal agreements failed in Leadville as they generally did throughout American industry.[54]

Although mining and smelting firms usually dealt directly with one another, ore buyers still performed an important service in the functioning of the camp. They brought producers and processors together, particularly when metallurgists needed mineral with certain amounts of iron, silica, lead, or limestone to prepare an optimum furnace charge. They also bought shipments outright in hope of selling them later at higher prices. And they served as arbitrators or "umpires" in disputes over the value of mineral shipped from mine to smelter. Because ore buyers sat astride developments in camp, they acquired a broad knowledge of the industry. It was no coincidence that some entered the smelting business. Meyer, Berdell, and Witherell were only the first to do so.[55]

Like all isolated mining regions, Leadville was a high-cost area, and this was clearly reflected in the costs of reduction. Laborers, furnacemen, smelters, and wheelmen earned from $2.50 to $6.00 a day, rates that ranged above the average pay in camp because smelting was usually unpleasant and sometimes dangerous. More than one worker was horribly burned by falling into a slag pot. The high price of labor was easily rivaled by the inflated cost of fuel. A bushel of charcoal fluctuated from ten to eighteen cents while a ton of metallurgical coke ranged from $10 to $23, an average figure being about $15.[56]

Yet smelting ores was a lucrative business for successful firms. If the fixed reduction charge of $25 together with penalties and returns on metal composed the total return, and if the costs of reduction averaged about $15, then the smelters enjoyed a profit margin of about $10 on every ton of mineral processed. If this estimate is accurate, the Grant enterprise earned about $250,000 in 1879, the La Plata $196,000, Billing and Eilers $107,000, and the Harrison works $81,000. Each would have been an excellent return on capital.

Such strong enterprises enjoyed profitable operations, but the demise of the small and weak came rapidly. The Leadville or Lizzie Smelter, along with Raymond, Sherman, and Mac-Kay, abandoned the business in short order; neither firm could overcome fuel shortages and what some observers regarded as poor management. Gage, Hagaman & Company kept its furnaces in blast for only two months before stopping because the plant could not get ores with enough lead to recover silver—an incredible revelation for early Leadville. The firm resumed operations for a short time in 1880, then passed from the scene. Poor furnace construction at the California plant forced the enterprise into frequent stoppages throughout 1879. The firm made repairs during the winter, but they failed to revive the company's fortunes; it failed the next year. The Little Chief and Adelaide mining companies also found it impossible to operate their smelters full time. After a year of frustration, both firms decided to dismantle their plants and concentrate on mining ores.

Some firms appeared to be successful yet still failed to survive. The Ohio & Missouri Smelting Company made a propitious beginning with one furnace, which ran steadily from May 1879 until the end of the year. The owners grew so confident of future prosperity that in January 1880 they installed a second reduction unit, but two months later the enterprise fell on hard times. The coke famine compelled the plant to cease operations, and this sealed the corporation's doom. It never reentered the industry. Another casualty was the old Malta Company, which succumbed in the spring of 1880. Neither the expertise of Franz Fohr nor the efforts of the Dixon family had managed to revitalize the region's original smelting firm.[57]

The frustrating saga of the Little Chief Mining Company may have typified the careers of the unsuccessful firms. This enterprise was financed and controlled by capital in Milwaukee and Chicago. Once in business the concern acquired several claims on Fryer Hill and hired George Hubbard Holt, a minority stockholder, to develop the properties. He opened several shafts and sold the ores to local works, but he and his colleagues decided to erect their own smelter in the belief that forward integration would increase profits.[58]

Holt began building the plant in March 1879. At first he made only slow progress because late winter snows and the tenuous transportation system impeded the shipment of materials to the upper Arkansas. Not until July did Holt finish the buildings and make final preparations to blow in the single blast furnace. At the last moment, however, he had to delay the start of operations because a crusher failed to arrive on schedule. As a temporary expedient he hired several men to break rock by hand, but this proved expensive, time-consuming, and inefficient. Once the crusher arrived, Holt again made preparations to begin, but this time operations were delayed when the Leadville Water Works notified him that it could not meet the smelter's needs. Not until the end of July did Holt resolve this problem. Once he got his furnace in blast, he produced bullion until September, when the plant exhausted its supply of coke and had to shut down. By this time Holt had fallen out with other company officials, who carped at his management—perhaps with some justification. This prompted him to sell his interest in the firm and resign. The new regime smelted more ores that fall, but operations remained sporadic and ceased entirely in 1880. With the furnace sinking into the foundation on account of poor construction, the enterprise finally decided to dismantle the plant and ship its mineral to custom smelters.[59]

The brief life span of the Little Chief smelter illustrated the obstacles to success. Poor planning, inexperienced management, faulty construction, and bad luck plagued the firm. It made no effort to hire a professional metallurgist, preferring to leave all technical matters to Holt, a man of limited knowledge in the field. But, in fairness to Holt, he had to double as the engineer in charge of the mine. As an integrated mining and smelting company, the Little Chief enterprise was somewhat unusual in the Leadville industry, but it provided another example of the difficulties faced by mining companies that integrated their operations forward into ore reduction.

Attrition took its toll. By the middle of 1880, more than half the firms that entered the business had failed, even as mineral production soared to new heights. Only seven smelting companies now remained. Of these the Grant enterprise continued as the largest, followed by the La Plata, Billing and Eilers, Harrison, Cummings and Finn, Elgin, and American firms. Their plants were large and efficient, their output huge.[60]

Yet this was not all, for the preeminence of Leadville also extended to technology. Eilers, Hahn, and other metallurgists who had come to the North Arkansas pushed the art and science of silver-lead reduction to its most advanced stage. They drew upon the successes and failures, experiments, and critical articles of the past decade. They used the best plant designs, employed water jackets, built dust chambers, controlled the composition of slag, and cut the losses of metal to relatively low figures. Although the carbonate ores were easier to process than silver-lead minerals found elsewhere, this in no way detracted from the industry's achievement. And this achievement culminated in a landmark study of smelting technology.[61]

In 1879 the United States Geological Survey, under its first director, Clarence King, decided to make a systematic investigation of three western mining districts. After some deliberation, King and his colleagues chose Leadville as well as Eureka and Virginia City, Nevada. The task of supervising the Colorado inquiry fell to Samuel F. Emmons, head of the Survey's Rocky Mountain division, which had its headquarters in Denver. A graduate of Harvard and later a student at both the Ecole des Mines and the Bergakademie, Emmons had the proper qualification, and in pursuing this research he had the foresight to hire able assistants, notably Whitman Cross and A. W. Hillebrand. Emmons, however, still needed another man to devote attention to smelting technology.[62]

The search for a competent metallurgist proved frustrating because such men's services were in great demand. Both King and Emmons wanted to hire Anton Eilers, who was adding luster to his reputation by his success in Leadville. But government regulations prohibited the survey from employing someone engaged in private business; and so the agency lost the opportunity to secure the premier engineer-entrepreneur in camp. By the spring of 1880, with the study of mining well under way, Emmons still had not found anyone he thought able to do the work on the smelters. Then in May when he wrote his immediate superior George F. Becker, he mentioned almost parenthetically that he was about to interview a Frenchman.[63]

This man was Antony Guyard, whose career in the minerals industry had spanned three continents. He had attended the Ecole des Mines, but he lacked the money to complete his course of study. Later he worked as a chemist for a manufacturer in London, a nitrate company in Peru, and a copper smelter in Normandy. How and why he came to Colorado remains unknown, but in May 1880 he applied for a position with the survey.

At their first meeting Emmons was not impressed, because Guyard "blew his own horn" and moved ponderously like an elephant. Yet Emmons recognized ability and soon rationalized Guyard's personality as a "national peculiarity" of Frenchmen. While the survey decided on his application, Guyard obtained employment as an assayer in one of the smelters. Emmons's superiors, still disappointed because they could not engage Eilers, vacillated on appointing Guyard because he had little experience in smelting silver-lead ores. They had Emmons interview more candidates, but none proved satisfactory. Emmons asked Eilers to suggest someone, but he could not strongly recommend anyone. Finally, the lack of acceptable alternatives and Guyard's avowed interest in the work made Emmons, Becker, and King appoint him to the staff.[64]

Guyard began his investigation in August 1880. Following Emmons's instructions, he visited every smelter in camp except the Little Chief and Lizzie, which by then had been dismantled. Somewhat surprisingly, he found owners and personnel very cooperative, particularly Billing and Eilers, August Werner of the Elgin Company, and James Brierton of the Harrison Works. Guyard focused on the chemistry of smelting and the composition of furnace products, but he also gathered information on assay values, furnace charges, reduction costs, bullion content, transportation expenses, refining prices, metal losses, and capital investment. And he garnered samples of ore and slag that he took to the survey's laboratory in Denver for analysis that winter.[65]

After writing a preliminary report, Guyard returned to France. Over the next few years he corresponded with Emmons and other members of the survey as the great monograph on Leadville took shape. Drafts of the metallurgical section made trans-Atlantic voyages. But when the final copy was nearly ready Guyard died suddenly in Paris on March 29, 1884. This was a serious blow to Emmons, who had developed a close friendship with the Frenchman. Hillebrand and Emmons made the final revisions, however, and when the massive *Geology and Mining Industry of Leadville* appeared in 1886 Guyard's research accounted for nearly a fourth of the study. While it superbly complemented Emmons's work on mineralogy and mining, Guyard's investigation was the most complete assessment of American smelting practices up to that time. Yet, like most scientific papers, by the time it made its way into print the technology of ore reduction had advanced so rapidly that Guyard's findings were obsolete.[66]

Nonetheless, in the brief span of three years a handful of entrepreneurs had created the largest, most important silver-lead smelting center in North America. They had done it in virtual wilderness, without the benefit of railroads, and without adequate supplies of fuel. Within Colorado the value of bullion shipped annually by plants on the upper Arkansas overwhelmed the production of the Boston and Colorado Company, which had dominated reduction and refining for a decade. These changes signaled a shift in the minerals industry from one oriented toward the eastern slope and South Park to one situated in the less accessible central Rockies.

Chapter 5

The Crest of the Continent

THE GREAT LEADVILLE BONANZA ROSE TO NEW HEIGHTS DURING the early 1880s. The mines on Fryer, Breece, Iron, Yankee, and Carbonate hills remained in operation, shipping enormous quantities of silver-lead ores to the smelters. A few older properties played out, but mines like the Evening Star, Iron Silver, and Robert E. Lee continued their steady production, and others like the Robert Emmet, Colonel Sellers, Little Johnnie, Henriett, Maid of Erin, and Wolftone sent forth new riches. Labor unrest, stock manipulation, and the collapse of the Little Pittsburg and Chrysolite companies detracted from the luster of the bloom but had little deleterious effect on the overall level of prosperity. Output rose from $11,300,000 in 1879 to $14,900,000 in 1880, and then, after a small decline the next year, to peaks of about $15,250,000 in 1882 and 1883. Silver production alone surpassed the value of precious metals mined in every other state and territory except California. Leadville was the greatest mining camp in the United States. And its smelting industry was likewise the greatest.[1]

The largest smelter in camp began the new decade with a change in ownership. On January 1, 1880, Judge James Grant sold his interest in the Grant works to Edward Eddy and William H. James, the ore buyers with whom the firm had long worked closely. In the years ahead the judge remained affiliated with the concern and did some of its legal work, but he had little to do with either management or decision-making. Meanwhile, James B. Grant and his new partners reorganized the firm as the Grant Smelting Company, while Grant himself purchased an interest in the sampling agency, known henceforth as Eddy, James & Company. The two enterprises did not merge, but the common ownership created an informal degree of integration. For all practical purposes these firms were one.[2]

After creating the new partnerships, Eddy, James, and Grant remodeled and expanded the smelter. When they began, the plant had eight furnaces with a capacity of about 160 tons of ore daily. During the next two years the firm shut down its units one by one and replaced them with larger models, all based on the rectangular Raschette design. By the end of 1881 the works had seven furnaces that could process 350 tons of ore a day. This was by far the largest in Leadville. The increased capacity permitted the enterprise to lower reduction costs and take advantage of economies of scale. This was important because the Leadville ore market remained highly competitive, and many mines were beginning to ship lower grades of carbonates. The arrival of the Denver & Rio Grande and the Denver South Park & Pacific railroads had also opened the possibility of sending large quantities of ore to smelters on the Great Plains—or in the valley, as people said—where reduction costs were lower.[3]

Grant was the great figure in the Leadville industry. Despite the long experience of Billing, Eilers, Hahn, Harrison, and others, many thought his plant was the best managed in camp, an accolade that also reflected the skills of Eddy, James, and Malvern W. Iles. The enterprise paid mining and coking firms promptly. No one ever protested its checks for nonpayment. And its ore bins always bulged with mineral. Ores flowed in from the Glass Pendery, Little Chief, Robert E. Lee, Evening Star, Highland Chief, Breece, and other mines. Silver-lead bullion moved steadily to Omaha. And when Iles discovered the metal vanadium in mineral shipped from the Evening Star and Aetna mines, the enterprise offered specimens for sale. They contributed a negligible amount to profits, but Iles correctly thought that vanadium was destined to play an important role in what he termed "the arts." Grant's sales foreshadowed the day when the industry would market a large array of metals besides silver and lead.[4]

The smelter produced enormous quantities of bullion during the first two years of the decade. From the $2,430,000 shipped to Omaha in 1879, output rose to $4,010,000 in 1880, but dipped to $3,080,000 in 1881 when Leadville's production declined. This was about one-quarter of the camp's shipments and represented the largest market share of any smelter in town.[5]

As the enterprise prospered, however, Grant and his uncle found themselves sued for patent infringement. In October 1880 Albert Arents, acting for himself and Winfield Scott Keyes, filed a suit claiming that the Grants had employed the famous siphon-tap invented at Eureka, Nevada, without paying the proper royalties for its use. Arents and Keyes petitioned the United States circuit court in Denver to award them $9,000 in profits foregone and another $27,000 in damages.

The complaint set the stage for a trial involving nearly every prominent metallurgist in the United States and Europe. The case hinged on the description of a crude tapping device discussed by John Percy in the 1870 edition of his *Metallurgy*, as well as the interpretation and translation of several German treatises. When the suit came to trial in 1882, Judge Grant headed a team of lawyers who contended that the siphon-tap was not unique but was similar to other devices used for many years. One side or the other obtained testimony from Grant, Eilers, Hahn, Arents, Keyes, Beeger, Fohr, Loescher, and others, as well as affidavits from Percy and several German metallurgists. Most testified that their firms had paid royalties for use of the siphon-tap, but the jury still found in favor of the Grants.

Dismayed but undeterred by the verdict, Arents and Keyes

took their case to the United States Supreme Court. Another three years passed, but in October 1885 the justices declared a mistrial. Rather than prolong the suit, the contestants decided to settle out of court. The Grants paid Arents and Keyes $2,500 and assumed the costs incurred in prosecuting the suit.[6]

In addition to smelting ores to bullion, Eddy, James, and Grant invested heavily in mining properties. With Horace Tabor and a group of New Yorkers, they organized the Hibernia Consolidated Mining Company, which took over the Surprise, May Queen, and Hibernia claims south of the Robert E. Lee mine on Fryer Hill. Eddy, James, and Grant also invested in the Small Hopes Mining Company, which owned the Forest City and Robert Emmet mines on Yankee Hill. Despite the modest assumptions of its name, the enterprise produced huge quantities of rich ore and paid large dividends to its stockholders. Grant himself bought a large block of shares in the Silver Cord Combination Mining Company, and for a time he managed the Highland Chief property. Ores from these firms did not always go to the Grant works for reduction, but the large stockholdings of Eddy, James, and Grant always assured the smelting firm of favorable consideration in contract talks.[7]

Operations at the smelter ran smoothly until the fateful night of May 24, 1882. Then, as a furnace crew drew off slag from the number three unit, waste material clogged the effluent system. Pressure in the furnace rose rapidly and blew out the tuyeres. As workmen ran for their lives, the molten metals set fire to the woodwork. The flames spread rapidly and engulfed the entire works. Before the conflagration could be extinguished, the plant burned to the ground, a total loss. James estimated the damage at $400,000, only a small portion of which was covered by insurance.[8]

The disaster threw Eddy, James, and Grant into a predicament. On the charred grounds of what had been Leadville's finest smelter lay huge quantities of ore already purchased. Beyond this, the firm had to fulfill other contracts with mineowners and suppliers of charcoal and coke. Eddy, James, and Grant soon decided to relocate and rebuild in Denver, but what were they to do before they could open a new plant? There was only one alternative, for it was essential to conserve capital and maintain some sort of cash flow. They decided to lease the Elgin smelter, which by then had fallen on hard times. Although the plant had a small capacity, it enabled the partners to maintain operations and honor their other contracts. By the time Eddy, James, and Grant gave up their lease in December, the Elgin smelter produced a respectable $580,000 in bullion, though this hardly compared with the $1,900,000 shipped by the Grant works before the fire. The combined output of $2,480,000 was a good showing in a difficult time. Eddy, James, and Grant had a bright future when they opened their new plant, but Denver's gain was Leadville's loss.[9]

Like Eddy, James, and Grant, Gustav Billing and Anton Eilers also expanded and remodeled their plant. In June 1880, after running steadily for nearly a year, they shut down their two furnaces to make repairs and delay bullion shipments until the railroads reached Leadville. During the next winter Billing and Eilers enlarged the plant by installing their third and fourth smelting units. They also replaced their oil and gas illumination with the crude electric lights of the day, and they added dust chambers to collect silver and lead volatilized in the smelting process. As a result, Billing and Eilers increased their production from $1,080,000 in 1879 to $2,100,000 in 1880 and to $2,500,000 in 1881. In the last two years this represented about 18 percent of Leadville's output, and the smelter surpassed the La Plata works as the second largest producer in camp. When Grant's works burned in 1882, the plant became the premier smelter in Leadville, a position it never relinquished.[10]

During this time Billing remained in the background conducting financial affairs, but Eilers involved himself in a wide range of activities. He served as an informal, unpaid consultant to Emmons on the great study of the geology and metallurgy of Leadville. On a more remunerative basis, he worked as a consultant to the Iron Silver Mining Company and for a short time in 1881 took over as superintendent. When Rossiter Raymond became president of the troubled Chrysolite mining company after the stockholders deposed the old management, he once again engaged his friend and former employee. Eilers no longer published articles on metallurgy, but from time to time he wrote public letters commenting on the work of others, and he remained active in the American Institute of Mining Engineers.[11]

After three successful years in the industry, the Utah smelter changed hands. On January 1, 1882, Billing and Eilers dissolved their partnership—why, no one knows—but in the settlement Billing acquired all the property and assets. Eilers received an unknown amount of cash. During the next few months Billing negotiated with August R. Meyer and other industrialists in Leadville, Kansas City, and St. Louis, the talks culminating on April 10 with the formation of the Arkansas Valley Smelting Company. This enterprise acquired the Utah works and Meyer's sampling agency so that from the outset it enjoyed the advantages of vertical integration. From then on the Utah plant was known as the Arkansas Valley smelter, or in later years simply as the AV works.[12]

The founders of the enterprise were men well known in mining and smelting circles. In addition to Meyer and Billing, the original board of directors included Arthur H. Meyer, Charles T. Limberg, and Horace Tabor of Leadville; Frank L. Underwood, Edwin E. Wilson, and A. W. Armour of Kansas City; and Theodore Plate and John Harrison of St. Louis. They set the capitalization at $500,000, divided into 50,000 shares of stock, each worth ten dollars at par. The directors elected Underwood president even though his 9,000 shares ranked second to the 11,000 owned by vice-president, August Meyer, who quickly emerged as the leading spirit in the venture. Billing acquired 8,000 shares, Tabor and Limberg 5,000 each, and the others the remaining 12,000. Over the next few years August Meyer purchased more stock and replaced Underwood as president. Billing, however, sold his interest in the firm and went to Socorro, New Mexico, to launch another smelting enterprise. Tabor also disposed of his shares, another fateful mistake on his road from riches to rags.[13]

The Arkansas Valley Smelter. William H. Jackson, photographer. Colorado Historical Society, Denver.

Only six weeks after the enterprise came into being, fire destroyed the Grant works, and so almost from the outset the Arkansas Valley Company owned the largest smelter in Leadville. Nevertheless, Underwood and Meyer moved quickly to enhance their position within the industry. Before the end of the year they installed two more blast furnaces, which brought the ore capacity to about 250 tons daily. The plant operated around the clock and sent thirty-three tons of silver-lead bullion to Meyer's refinery in Kansas City every day. Yet in spite of the increase in size the production of $2,500,000 in 1882 only equaled the value of bullion shipped by Billing and Eilers a year earlier. Ore grades were falling.[14]

The next year proved far more difficult for the enterprise. Because of a poor spring runoff, the firm found water in short supply and had to share what it had with the American Mining and Smelting Company. Meyer and his colleagues later constructed their own reservoir to avert similar situations in the future, but of course this did little good in 1883. The prices of silver and lead also declined, curtailing production from mines in camp. Yet the most serious problem was that Meyer found he could not secure enough high-grade lead ores to mix with low-lead minerals in making reduction charges. This was the harbinger of a grave predicament that eventually altered the industry's structure. Part of the problem stemmed from the appearance of new competitors: the Grant works at Denver, Eilers's new Colorado Smelting Company, and the Pueblo Smelting and Refining Company, all of which operated in the valley, where costs were lower. Beset by a plethora of problems, the output of the Arkansas Valley smelter fell to $2,200,000 in 1883.[15]

The La Plata Mining and Smelting Company went through a series of changes similar to those of its rivals. Early in 1880 the firm had a change in management when Nathaniel Witherell replaced Charles B. Rustin as president, and Theodore Berdell took over as vice-president. Throughout the year they operated four blast furnaces that had an aggregate capacity of 120 tons daily, but in 1881 they followed the general trend in the industry by installing two enlarged units that permitted the plant to process 200 tons of mineral a day. They also built dust chambers to capture bullion that would otherwise have escaped with the furnace fumes.[16]

Witherell and Berdell also continued to develop their mining properties. As a rule, they produced ores that were low in silver content but high in lead, although they hoped to locate richer deposits of silver. Nonetheless, they regulated output so as to coincide with the need for base metal in preparing reduction charges. This measure of vertical integration guaranteed the firm adequate supplies of lead to mix with ores shipped by mines like the Robert E. Lee, Evening Star, and Silver Cord Combination. And the La Plata concern became one of the first companies in town to purchase ore in other camps when in 1880 it bought several large consignments from the Robinson Consolidated Mining Company, whose smelter at Kokomo, north of Leadville, had shut down because of a dearth of lead.[17]

The La Plata firm also established a unique social system, quite unlike anything devised by rival companies. For their employees Witherell and his associates erected a group of dwellings that clustered around the plant "like a Swiss village about its public works," according to one admirer. The enterprise provided its men with a library, reading room, and billiard parlor for use after hours and also offered some medical services. These amenities were unusual for the time, but this early form of welfare capitalism was not entirely altruistic, for the firm hoped to attract better laborers than its competitors, maintain a steady work force, and keep its two hundred workmen under steady supervision. The corporation also displayed the hard side typical of American industry in that day. One smelter worker who had been severly burned by molten slag had to fight a long legal battle to get some compensation for his injuries; his success in court was unusual for the Gilded Age.[18]

Under the leadership of Witherell and Berdell, the La Plata firm increased production during the early eighties. From the $1,700,000 shipped in 1879, output rose to $2,300,000 in 1880; declined to about $2,000,000 the following year—when Leadville's production slipped—and then increased to an all-time high of $2,400,000 in 1882. From the large profits that flowed into the treasury from sales of bullion, Witherell and his colleagues declared a total of $610,000 in dividends, the last coming in September 1882. After that, financial problems and industry wide difficulties beset the corporation and altered its operations.[19]

As these firms pursued their separate but intertwined destinies, the St. Louis Smelting and Refining Company, which ran the oldest plant in camp, found itself confronted by a host of problems. The first involved management. In 1880 Edwin Harrison and his colleagues decided to replace George B. Hynes, who had long been unhappy in the carbonate camp. The search for a successor was brief. Harrison soon settled on the far abler, though crusty, Franz Fohr, who was seeking new employment after the final collapse of the Malta Company. In the first year he had charge of the smelter, Fohr shipped more than $900,000 worth of bullion, but this was far below the production of other plants. What hurt output was a dreadful November fire that destroyed the works in Missouri and compelled Harrison to sell bullion produced in Leadville to other refiners until he could rebuild.

Harrison and Fohr also recognized that the smelter had done poorly because it was obsolescent. While rebuilding the plant in St. Louis, Harrison had the firm appropriate enough capital so that Fohr could remodel. As money arrived from St. Louis, he curtailed operations, did away with the old smelting units, and replaced them with four large furnaces that increased the daily ore capacity to two hundred tons. He also installed dust chambers and other machinery and built a new sampling agency to replace the one sold to August R. Meyer.[20]

With renovations complete, Fohr blew in the new furnaces and resumed full-scale operations in September 1881. Smelting went smoothly for the rest of the year, and the plant shipped about $340,000 in bullion, nearly all of it during the last quarter. Fohr kept the furnaces in steady operation throughout the next year, and output rose to a new high of $2,000,000. But bad luck continued. On the evening of May 17, 1883, nearly a year to the day from the time fire destroyed the Grant smelter, a blaze broke out in the Harrison Works and caused an estimated $60,000 to $70,000 in damage before it could be extinguished. The plant was not a total loss, however; so, unlike Eddy, James, and Grant, Edwin Harrison and his associates decided to continue operations in Leadville. After rushing repairs during the next few weeks, Fohr set two furnaces in blast at the end of June and blew in the other two in early July. After that he experienced some difficulty in obtaining carbonate ores but still managed to continue operations. By the end of the year he had shipped $1,500,000 worth of bullion.[21]

The major smelting firms enjoyed a good level of prosperity during the early 1880s, but dark clouds appeared in the latter half of 1883. By this time the bountiful supplies of silver-lead carbonates, so easy to smelt into bullion, had begun to give out. Mining companies tunneled farther and farther into the earth in search of pay ore, but in the lower workings they found minerals richer in sulfur and poorer in silver and lead. The carbonate shortages experienced by the Arkansas Valley and Harrison plants late in 1883 were the first signs of a new era.

The appearance of dry ores—those with little lead—put a greater premium on higher grades of mineral, just when the rise in sulfide production demanded the installation of roasting furnaces to give the ores a preliminary treatment. This interrelated situation meant an increase in reduction costs. Over the next few years the industry responded to the change by increasing the breadth of its purchasing area, smelting charges with the minimum amount of lead, and integrating operations backward by building roasting furnaces. In the years to come, the preliminary treatment of sulfides grew so precise, sophisticated, and important that it virtually emerged as an independent branch of metallurgy.

But too much sulfur and too little lead were only two problems in a Pandora's box of troubles opening for the Leadville industry. Just as ominous was the rise of the valley smelters that had begun to challenge the virtual monopoly the Leadville plants had held since 1877. What had opened the competition were the railroads—the Denver & Rio Grande, which connected Pueblo to Leadville in 1880, and the Denver South Park & Pacific, which linked Denver to Leadville in the same year. They permitted miners to ship ores to the plains, where costs were lower than in the central Rockies. Shrewd entrepreneurs like Anton Eilers, James B. Grant, and the men behind the Pueblo Smelting and Refining Company had seen this early. By 1883 their firms had entered the Leadville ore market.[22]

Despite the growing competition, the smelters still tried to act in concert for their mutual benefit. As early as 1880 firms in Leadville and the Pueblo Smelting and Refining Company had agreed to boost reduction fees on all contracts expiring at the end of the year. This covenant, like most pooling arrangements, fell apart in a few months; but two years later the smelting firms tried again, this time joined by the new companies of Grant and Eilers. On October 11, 1883, the smelters issued a joint statement that specified minimum percentages of lead, a standard return on fractional ounces of silver and gold, and uniform procedures for performing assays, umpires, and moisture determinations. The communique created a furor in mining circles, and a few shippers, notably the Silver Cord

Combination, closed down in protest. Yet the October compact, like its predecessor, soon fell apart, this time when Grant and his associates withdrew.[23]

The declining price of lead also contributed to at least a measure of concerted action by miners and smelters. During the summer and fall of 1883, trade journals predicted that the metal would decline in value sometime in the next year, prognostications that became self-fulfilling. Rather than wait for a potential decline, several eastern refiners liquidated their inventories, causing the price to fall sharply by December 1883. The plunge forced several important mining companies in Colorado to close, and this in turn put pressure on the smelting industry, which had to have lead to collect silver. In response, a group of miners and smelters met in Denver to see what steps might be taken to bolster the situation until prices rose. The meeting ended in the formation of the Western Mining and Smelting Association, which numbered among its members such prominent individuals as Edward Eddy of the Grant enterprise, Henry I. Higgins of the American Mining and Smelting Company, Charles Limberg of the Arkansas Valley firm, and Henry Coom of the La Plata concern. Among the mineowners were William H. Stevens of the Iron Silver and Tingley S. Wood of the Silver Cord.

The association hoped to raise the price of lead, but when it realized it could do little about market values it turned its attention to railroad rates. The group charged that freight tariffs from Leadville to Denver and from there to eastern cities were excessively high. That was not an uncommon assertion, but the association went further, claiming that certain favored smelters received rebates. There was some truth in this, but the organization could do nothing to eliminate the practice. Everyone wanted rebates anyway. The association did persuade the lines to make a few concessions on ore and bullion shipments, but they had no effect on the price of lead, which remained low for more than a year, long after the association had dissolved. Because the price remained low, some smelters in Leadville curtailed operations. And some even closed their doors.[24]

In contending with such difficulties, the Arkansas Valley Company used new management and new capital to hold its position within the industry. About the time that August R. Meyer became president of the firm, he relocated in Kansas City, Missouri, to manage the enterprise, his mining investments, and his new Kansas City Smelting and Refining Company from there. He had his old friend and associate Charles T. Limberg appointed manager of the Arkansas Valley works. Once on the job as of January 1, 1884, Limberg installed matte taps in all the furnaces to aid the treatment of sulfide ores. For the first time the plant shipped small quantities of matte in addition to bullion. When Limberg found he had excess capacity at the sampling works, Meyer negotiated a contract with the Pueblo Smelting and Refining Company. Under this arrangement the Leadville plant shipped more than $400,000 worth of mineral down the valley to its rival that year. Meyer and Limberg proved themselves skillful managers. Drawing on their financial resources, they secured enough supplies of mineral to operate steadily throughout 1884, and the Arkansas Valley works shipped bullion and matte valued at nearly $2,400,000.[25]

Despite the good showing, Meyer and Limberg encountered a far more serious problem the next year. Once their contract with the Pueblo Company expired, this enterprise entered the Leadville ore market itself, offered shippers very lucrative returns, and virtually cornered the production of minerals high in lead. Like other smeltermen in camp, Meyer and Limberg found they could not meet such prices and still operate profitably. Limberg shut down one furnace after another. By May 1885 he had only two units in blast, and these operated with ores purchased at a loss. The situation was hardly satisfactory, and it could not continue for long if the firm intended to remain in business.[26]

In August, Meyer and two directors arrived in Leadville to look at the problem firsthand. With Limberg, they discussed the possibility of relocating the smelter on the plains at either Denver or Pueblo. But after weighing the alternatives, Meyer and his associates decided to continue operations and do whatever was necessary to restore the profitability of the largest plant in Leadville. To this end the firm suspended all operations in September so that Meyer and Limberg could renovate the works, install larger and more efficient furnaces, and acquire an adequate supply of ores rich in lead. The work lasted several weeks. When Limberg blew in seven new furnaces in October, the plant had increased its ore capacity from 250 to 300 tons daily.[27]

Bullion production during this time reflected Meyer's problems. Output in 1885 fell to about $1,300,000 roughly half the amount shipped to refineries the year before. The decline also cost the plant its accustomed position as Leadville's leading smelter, although this proved to be only a temporary aberration. In 1886, when business conditions allowed the plant to run all year, the AV works smelted ores worth $2,300,000— once again the largest output in camp. Yet this figure failed to equal the shipments made in 1881 and 1882, even though it came from a much larger quantity of ores. Lower prices for silver and lead accounted for some of the fall, but, clearly, the grade of mineral mined in Leadville had declined.[28]

Whereas the Arkansas Valley firm surmounted this series of difficulties, the La Plata Mining and Smelting Company fell into financial quagmires, if not financial chicanery. By the end of 1882 Nathaniel Witherell and his associates had enjoyed several years of high profits and large dividends, but about this time unsettling rumors about internal finances appeared in mining journals. Speculators manipulated the firm's securities on the New York Stock Exchange. Then, at the end of the first quarter of 1883, Nathaniel Witherell reported a profit of only $20,500, far less than usual, and he conceded that the officers had taken out loans unbeknown to the stockholders. In explaining this action, Witherell contended that borrowing money was a normal practice in the smelting business, that the managers had guaranteed the loans personally, and that they expected to reduce the debt from projected earnings in future weeks.

Witherell's explanation, however, failed to convince a large number of stockholders in London—where the controlling interest resided—and they formed a committee to investigate the American management. Under the direction of William Abbott, the group examined the books and made its report at a company meeting in May. Abbott pointed out that the enterprise had unjustifiably paid dividends throughout 1882 and had incurred a debt of £38,000 in doing so. Witherell and his associates had reduced the figure to £25,000, but Abbott urged that the firm be reorganized as an English company. A majority of the stockholders agreed, and that summer the assets and liabilities passed into the hands of a new enterprise. Witherell and his associates were so discredited, however, that they sold what shares they had and left the concern.[29]

The new La Plata Mining and Smelting Company, Ltd., engaged Henry Coom, an experienced mining engineer, to take charge of operations in Leadville; but no sooner had he done so than he had to confront a host of problems. By 1883 the La Plata smelter was the oldest in camp, and because of its obsolescence and inefficiency the firm became the first enterprise in Leadville to suffer from the competition of the valley smelters and the declining price of lead. When Coom found he could not compete, he recommended an extensive renovation of the plant. The directors concurred and raised the necessary capital. In September Coom suspended operations and kept the furnaces out of blast for sixty-six days while he remodeled the works, installed new machinery, and purchased large stocks of ore rich in lead. During this time he maintained some cash flow by selling about a thousand tons of high-grade mineral to the Harrison Works. Then in December he resumed operations.[30]

Despite the improvements, Coom encountered what must have seemed like an unending series of troubles. When the declining price of lead compelled mining companies to curtail production, he found himself unable to get enough mineral high in base metal to mix with dry ores. He cut back operations, and the smelter limped along at partial capacity until the end of March 1884, when he took all his furnaces out of blast. He resumed operations two months later in June, but with lead-bearing ores purchased at a loss. This was hardly satisfactory, but even more ominous was the revelation that the firm's mining costs were far higher than average. What was even worse, these properties were at last playing out.[31]

In an effort to get more lead for the La Plata furnaces, the directors took a novel step. During the winter they decided to build a refinery to desilverize their bullion in Leadville rather than sending it to eastern works. They hoped to obtain enough base metal to mix with dry ores, even though recycling would deprive them of a marketable product. Future profits—if any—would have to come from silver alone. The directors authorized a sale of debentures to finance the plant, and construction went forward in the spring of 1885. But, before the project could be completed, Henry Coom died in Denver on May 20.[32]

Although Coom's death was a setback, the firm soon replaced him with Reuben Rickard, a veteran metallurgist experienced at Eureka, Nevada, among other places, and a member of the well-known Rickard family of mining engineers. He assumed responsibility for the overall management in Leadville, but the actual work on the refinery fell to L. L. Humphreys, a metallurgist the La Plata firm had lured away from Swansea, Wales. Encouraged by a small profit in the second quarter, Rickard and Humphreys pushed the project to completion and commenced operations in September. They enjoyed technical success, but the refinery proved unprofitable and had to be abandoned. During this time Rickard continued smelting silver-lead ores, but he could never run the plant at full capacity. The works limped along through the next winter and on until May 1886, when unusually bad roads prevented delivery of ore to the smelter and Rickard decided to shut down entirely.[33]

The work stoppage created consternation among the debenture holders. In June one man filed a petition in London calling for the enterprise to wind up operations and liquidate its assets. The directors naturally thought this premature, and to counter such a move they agreed to a second reorganization. The creditors acquiesced, and the property soon passed into the hands of a new corporation appropriately, though not imaginatively, named the New La Plata Mining and Smelting Company, Ltd. Yet the stockholders had to pay a heavy price to keep going. Under the arrangement worked out with the debenture holders, the firm had to assess the shareowners $240,000 to pay off debts and provide new working capital. With an infusion of money, Rickard negotiated contracts with several mining companies and resumed operations, though, as usual, not at full capacity. But during the winter, as the snows on the North Arkansas blanketed the city of Leadville, rumors circulated that the La Plata firm was ready to go out of business.[34]

In the spring of 1887 the directors of the company, no longer willing to rely on reports from the local management, decided to have an outsider make an independent assessment of the smelter's prospects. To do the investigation, they engaged Philip Argall, a portly young British mining engineer soon to make a world reputation for himself in the milling business. He arrived in Leadville that summer, took one look around the smelter—so he later said—and cabled his blunt opinion to London. He told the directors that they should either appropriate enough capital to install modern equipment throughout the works or else cease operations entirely. Argall's cable brought matters to a head. Having lost heavily over the past few years, the directors voted to liquidate the corporation. And so another smelter in Leadville departed the industry, a casualty of obsolescent technology, inadequate capital, poor location, and perhaps bad management.[35]

The Harrison Reduction Works found itself under the same pressures as its rivals. After repairing the damage caused by the fire of May 1883, manager Franz Fohr resumed operations, only to notice large quantities of sulfides coming from mines in California Gulch, particularly the Colonel Sellers, with which the smelter had a year's contract. He also learned that he could not get enough lead ores to prepare furnace charges. New contracts between the valley smelters and the Silver Cord Combination, Evening Star, and other major suppliers had tied up production. The shortage of lead grew so acute that Fohr had to close down in August. This prompted Harrison to have the enterprise appropriate capital so that Fohr could install roasting kilns to burn off a portion of the sulfur before reduction. As the units neared completion, Fohr opened negotiations with lead producers, acquired sufficient quantities to resume operations, and blew in one furnace after another during November. To one inquirer he professed that he was no longer worried about securing adequate supplies of lead, but this statement was either a gross miscalculation or sheer bravado. During the winter Fohr rarely operated the plant at capacity, even though his ore bins reportedly bulged with twelve thousand tons of silver-bearing rock.[36]

Shortages of lead plagued operations for the next two years. For a time Fohr kept the furnaces in blast by resmelting his bullion with dry ores. He paid premier prices for whatever shipments of high-lead mineral he could get from the Adams, Henriett, and Lillian mines. He also had metal-bearing rock purchased in the distant San Juan shipped to Leadville despite the huge transportation costs incurred by the roundabout routing. Such expedients were hardly satisfactory to Harrison or anyone else, but there was no alternative except to close down. Fohr rarely operated the works at capacity until the latter half of 1885, when the Pueblo Company lost its grip on the local market and other mines came into production. Only then did the St. Louis company increase its purchases of lead ores and run on a steadier basis.[37]

Fohr's output reflected his problem with ore supplies. Throughout the middle years of the decade, his production never approached the $2,000,000 achieved in 1882. After shipping bullion worth $1,500,000 in fire-shortened 1883, he smelted an equal value in 1884; but after that output slipped to $1,300,000 the next year and rose to only $1,400,000 in 1886. Lower prices for silver and lead accounted for some of the decline, but the major cause was a dearth of base metal that prevented Fohr from operating at capacity.[38]

In the meantime, one of the smaller enterprises formed in 1879 emerged as a major competitor in the Leadville ore market. This venture was the American Mining and Smelting Company, organized by Caleb B. Wick, Henry I. Higgins, and several men engaged in the railroad industry. Originally the corporation acquired the Little Ellen lode and erected a modest smelter that had two furnaces and a capacity of sixty tons of mineral daily. Like most integrated firms of this type, the American Company devoted the bulk of its capital to developing mines and left the smelter operating largely as an adjunct, despite the eminence of its metallurgist, Otto H. Hahn. The meager output of $250,000 in 1879 and $300,000 in 1880 reflected the limited goals of the original entrepreneurs.[39]

But then the enterprise shifted its objectives, perhaps in response to the success of Grant, or of Billing and Eilers, or perhaps because of the presence of Hahn. No one knows. What is important is that the company changed its operating role from miner of ores to processor of ores. The shift came under the leadership of Higgins, who during 1880 emerged as the leading spirit in the venture. Over the next two years he and Hahn enlarged the smelter by installing additional blast furnaces that gave the works a daily capacity of two hundred tons of mineral. Production rose sharply to $1,000,000 in 1881, $1,100,000 the next year, and $1,700,000 in 1883, figures that compared favorably with the Arkansas Valley, La Plata, and Harrison works.[40]

The mid-eighties proved difficult for the Leadville smelters —years no less difficult for the American plant—but 1884 proved particularly significant to the firm, for in February Higgins and his associates brought off one of the largest mergers seen in the industry up to that time. The winter months witnessed the culmination of negotiations with Abner B. Thomas and Barton Sewell of the Chicago Smelting and Refining Company, a venture that had processed ore and bullion for more than a decade. In a complicated transaction the Chicago enterprise acquired the American firm as a wholly owned subsidiary, but in the exchange of cash, stock, and corporate positions, Higgins became president of the parent company. At the same time the American enterprise purchased the Eureka mine at Gunnison, Colorado, and leased the works of the Royal Gorge Smelting Company, which had a small plant at Canon City, Colorado, reducing ores shipped from the Gunnison and San Juan country. Because of its position in the lower Arkansas Valley, this smelter eliminated the necessity of hauling

mineral uphill to Leadville for reduction. To finance these ac-
quisitions, the Chicago enterprise increased its capitalization
by $200,000. Higgins, Daniel P. Eells of the Commercial Na-
tional Bank of Cleveland, and Charles Otis of the Otis Iron and
Steel Company of Cleveland purchased most of the new stock
issue.[41]

Meanwhile, a new management took over the American
smelter in Leadville. Hahn left to accept a more lucrative position in Pueblo, and Higgins replaced him with Sidney Bretherton, who had worked at the Arkansas Valley smelter. Fred H.
Ketchum succeeded Wick as general manager. Together with
Higgins, they provided the leadership in Colorado during the
next few years.[42]

Despite the merger, the American smelter still had to contend with the problems facing the Leadville industry. Owing to
the decline in lead production, Higgins employed the Eureka
mine to augment supplies of base metal for both Colorado
smelters—in fact, the property may have been purchased for
just that purpose. He also opened a sampling agency in Aspen,
a camp located west across the mountains and just then approaching its boom years. Rumors suggested that Higgins
planned to erect a smelter there, but the reports proved false.
He did, however, follow the lead of the La Plata corporation
and built a desilverizing plant in Leadville to get more lead for
the blast furnaces. Still, output fell to $1,500,000 in 1884.[43]

The next year Higgins and Bretherton increased production
even though they found their operations hampered by shortages of lead. For a time they garnered the entire output of the
Iron Silver mine, although by contract they had to share a
portion of the ore with Leadville's other smelters. When this
provision ended in June, the plant received all shipments for
the rest of the year, although they turned out to be only about
one-third of what Bretherton had expected. At the same time,
as increasing amounts of sulfides came onto the Leadville
market, Higgins and company appropriated enough capital to
build several reverberatory furnaces for roasting the mineral
before reducing it to bullion. Bretherton installed the units
during the summer. Despite their efforts, however, Higgins

and Bretherton proved unable to operate the plant at capacity for much of the time because of the dearth of lead, and this was reflected in their production of $1,600,000, a meager $100,000 increase over what they had achieved the previous year.[44]

Better times came in 1886. At the outset Higgins leased the mines of the Iron Silver and Iron Hill Consolidated firms to ensure the works greater supplies of lead. By good luck the price also rose and brought other properties into production. Such an improvement came at an opportune time for Higgins and his associates, because on September 24 fire engulfed their refinery in Chicago, causing $15,000 to $20,000 in damages. Although the disaster compelled Higgins and Bretherton to modify operations in Leadville until they could rebuild, the American works enjoyed its best year to date. Bullion production rose to a new high of $2,100,000, a close second to the output of the Arkansas Valley plant.[45]

By this time, only the Arkansas Valley, American, and St. Louis firms remained as major producers of silver-lead bullion. They had survived through good management, ample financial resources, and judicious expansion, but they all suffered from competitive disadvantages. They drew ores from more limited markets than their rivals in the valley, and they had to pay more for fuel and flux, labor and materials, all of which translated into high reduction costs. Their position was tenuous indeed.

Perhaps because of their vulnerability, the three major smelting firms became involved in mergers. The American Mining and Smelting Company, already a subsidiary of the Chicago Smelting and Refining Company, found itself a subsidiary of another corporation when its parent firm joined with the Aurora Smelting and Refining Company of Illinois to form the Chicago and Aurora Smelting and Refining Company. The Harrison Works went the same way in 1890 when Harrison and his associates sold the controlling interest in the St. Louis company to a group of entrepreneurs headed by Albert Schneider and Justus Jungk, men well known in the Utah and New Mexico industries. They in turn transferred the enterprise to a syndicate organizing what became the National Lead

Company—the "White Lead Trust." The Harrison and American works unquestionably suffered from the competition of the valley smelters, but they also had to contend with the potent Arkansas Valley plant, by now a constituent of the great new giant in the industry, the Consolidated Kansas City Smelting and Refining Company.[46]

This firm came into being on April 1, 1887, in the course of a complex merger engineered by the hard-driving August R. Meyer. On that day the new corporation acquired the El Paso Smelting Company, the Kansas City Smelting and Refining Company, the Mexican Ore Company, and two-thirds of the stock in the Arkansas Valley Smelting Company. These ventures held plants and properties that made Meyer's corporation the most highly integrated, broad-based, and dominant firm in the industry at the moment of its inception. The new concern had smelters at Leadville, Colorado; El Paso, Texas; and Argentine, Kansas, as well as a refinery in Argentine, a company town Meyer had built outside Kansas City. Through the Mexican Ore Company, the Consolidated firm also acquired several mines in the Mexican states of Coahuila and Sierra Mojada, a network of offices throughout the country, and sampling agencies at El Paso, Laredo, and Eagle Pass, Texas, the three ports through which Mexican minerals entered the United States.

The Consolidated Kansas City enterprise had sprung to life through the work of men well known in mining and smelting circles. Its president, treasurer, and leading figure was Meyer, who had moved his residence from Leadville to Kansas City some time before. Holding the position of first vice-president was Nathaniel Witherell, who manned the New York offices at 20 Nassau Street along with Edward Brush, the assistant secretary. Serving as second vice-president was Robert S. Towne, a man whose shadowy figure belied his significance in the Mexican mining industry. A protégé of Meyer, Towne had learned the minerals business in Colorado before turning his energies to Mexico. It was Towne who had erected the smelter in El Paso and assembled the properties now belonging to the Mexican Ore Company. The firm's trustees—or directors—included

Meyer, Witherell, and the old Leadville man Theodore Berdell, as well as John Quincy Adams, representing Boston capital, and A. Foster Higgins of Spencer Trask & Company, an investment banking firm in New York. They and two other trustees set the capitalization at $2,000,000.[47]

From the outset Meyer and his associates sought to expand their plants and improve their competitive position within the industry, but this required more capital, and so raising money became one of Meyer's chief tasks. He gained access to financial markets in New York through Spencer Trask & Company and to those in Boston through the old but still prominent Adams family, which had helped found the original Kansas City enterprise. During the winter of 1888 Meyer used his personal contacts to obtain subscriptions for $333,000 worth of stock that remained unsold. Yet this failed to provide the funds needed by the enterprise, and in 1890 Meyer sought to raise $1,000,000. In this case the company sold gold bonds bearing six percent interest through a syndicate composed of Lee, Higginson & Company of Boston and their counterpart in New York, Kuhn, Loeb & Company. Meyer and his colleagues guaranteed the securities by placing a mortgage on their three smelters. Two years later the firm increased its nominal capitalization from $2,000,000 to $2,500,000 by distributing new shares to its owners in the form of a stock dividend.[48]

Part of the capital raised went into the expansion of the Arkansas Valley smelter, which remained a central constituent of a far-flung minerals empire. Meyer and his colleagues saw to it that their local managers received the money to install additional roasters, enlarge the blast furnaces, improve the dust chambers, and replace older machinery as it wore out. By the early nineties they had increased the ore capacity to 300,000 tons yearly—or 820 tons daily—the largest of Meyer's three smelters. Yet despite the massive volume of ore reduced to bullion, all of which was shipped to Argentine for refining, the Arkansas Valley works had the firm's highest unit reduction cost, $5.87 per ton in 1892.[49]

As in former times, the plant still drew its chief supplies of ore from the mines of central Colorado. Leadville remained the major source of mineral, but it was the emergence of Aspen that enabled Meyer and his associates—as well as their rivals in camp—to acquire what they needed to keep all their furnaces in blast. Still, the rising production of dry ores plagued operations and forced Meyer and his managers to rely on northern Idaho and northern Mexico to provide the lead essential in the smelting process. This meant that instead of reducing some ores in their natural market—El Paso or Argentine—Meyer's firm had to pay extra freight charges to haul the mineral from the plains up the mountains into the central Rockies, which is one reason the AV plant had higher reduction costs and why the valley smelters surpassed its output.[50]

Even as Meyer enlarged the AV works, the smelting industry in Leadville had changed forever. Eddy, James, and Grant had relocated on the Great Plains. Billing and Eilers had sold out. So had Cummings and Finn. The La Plata firm had gone out of business. The Elgin plant remained in operation, managed by the Manville Smelting Company, but its production was insignificant. New entrepreneurs were ready to try the pyritic smelting process, but their future was conjectural. Only the Arkansas Valley, Harrison, and American works continued as large processors; yet their output failed to maintain Leadville's preeminence in ore reduction. The leadership had passed to Denver and Pueblo.

Chapter 6

Smokestacks on the Plains

DURING THE SIXTIES AND SEVENTIES MANY HAD RECOGNIZED that the cities and towns on the Great Plains held excellent potential as ore reduction centers. Some noted that labor, fuel, and material costs were far less here than in the isolated mining camps of the high country. Others saw that an efficient system of railroads could unite a wide variety of minerals for treatment in smelting furnaces. Yet the absence of rail service had compelled the early entrepreneurs to locate their plants in the mountain fastness near the sources of fuel and ore. During the seventies, however, railroad construction began to link the major cities, towns, and mining camps and opened new opportunities to the most perceptive smeltermen.

Denver was the nexus of railroad building in Colorado. The construction of the Kansas Pacific and the Chicago, Burlington, and Quincy lines gave the city direct service to the East, and the Union Pacific road, through its subsidiaries, the Colorado Central and Denver Pacific, provided an east-west link through Cheyenne. Within Colorado the Denver & Rio Grande Railway built south from Denver to Colorado Springs, Pueblo, and Trinidad, then west from Pueblo to Leadville and Durango. West of Denver the Colorado Central Company laid narrow-gauge rails up Clear Creek Canyon to Central City, Idaho Springs, and Georgetown. And the Denver South Park and Pacific enterprise extended its tracks to Fairplay, Alma, and Leadville.[1]

As these roads made Denver a highly prospective site for the reduction industry, Nathaniel P. Hill and his colleagues in the Boston and Colorado Smelting Company had begun to think about consolidating their operations on the Great Plains. A decade of rapid growth had carried the Black Hawk plant to the limits of expansion in the narrow, high-walled canyon carved by North Clear Creek. Since 1876 production had risen slowly, partly because of the physical constraints imposed by the setting. Increases in the cost of fuel had eroded profit margins, and high freight tariffs had thwarted Hill's plans to switch from wood to coal. But moving the plant to a more centralized location offered Hill the prospect of further growth because the enterprise could draw on the production of a wider range of mining camps. Finally, the suit filed in circuit court at the behest of Carl Schurz forced the company's hand. Hill and his associates decided to consolidate their smelting and refining operations at the edge of the eastern slope. And it was to Denver, with its excellent railroad system and lower costs, that the enterprise looked for a new plant site.[2]

Relocation was no simple matter, however, because a peculiar track arrangement connected Denver to the mining camps on the forks of Clear Creek. Freight originating in Denver had to travel over the broad-gauge track of the Colorado Central Railroad as far as Golden. There yard crews transferred the cargo to narrow-gauge rolling stock for the journey into the mountains. The opposite procedure applied to shipments coming down from the mining towns. People in Denver and in the high country had long urged the Central to lay a third rail permitting direct, less-expensive transport between Denver and Golden, but William A. H. Loveland, president of the line, refused. He used various pretexts to justify his position, but the real reason for his opposition was that he feared his warehouses and retail stores in Golden might suffer if Denver instead of Golden became the principal transfer point.[3]

Hill thought the construction of a third rail was vital. If he erected a smelter in Denver, the bulk of his mineral would roll down Clear Creek Canyon on narrow-gauge ore cars as far as Golden, where it would have to be transferred to broad-gauge rolling stock for shipment to the works. Hill considered this expensive, time-consuming, and unnecessary. It defeated the purpose of relocation. In the summer of 1877 he had tried to persuade Loveland to lay a third rail and reduce the expense of shipping coal to Black Hawk, but Loveland had refused, realizing that the smelting company was dependent upon his line. Once Hill and his associates decided to shift their location, Hill reopened his talks with Loveland, who remained adamantly opposed to laying a third rail.[4]

Hill then shifted tactics. He let it be known that he intended to build his own railroad between Denver and the towns above Clear Creek. In retrospect it seems that this scheme was little more than a clever ploy to force the Central to come to terms, but it had to be taken seriously at the time. One commentator wrote that Loveland would have to reach an agreement with Hill because the railroad bore a heavy debt and would fail to meet its obligations if it had to compete with another carrier.[5]

Regardless, Hill wasted little time in putting pressure on the Colorado Central. In November 1877 he, Wolcott, and three other men organized the Denver and Rocky Mountain Railway Company. Popularly known as Hill's Road or the High Line, the enterprise proclaimed its intentions of laying track from Denver to Black Hawk, Central City, Georgetown, and Caribou, then across the state to the Utah border. In January a Denver newspaper reported that the directors of the smelting company had agreed to finance the project. Almost simultaneously, Hill purchased a wagon road between Idaho Springs and Georgetown for part of the route, set surveyors to work, and sent Wolcott to Boston to assist the eastern management in raising money.[6]

Businessmen at Central City and Georgetown were enthusiastic about Hill's Road, but their ecstasy was short-lived. In March, rumors appeared that Loveland had capitulated and the High Line would not be built. A spokesman for Hill denied the report as hearsay, but a Denver newspaper soon published a story claiming that Jay Gould and Sidney Dillon of the Union Pacific Railroad, which controlled the Colorado Central, had assured the smelting enterprise that Loveland's firm would make concessions for carrying ores to the new plant.

Events proved that the newspapermen were indeed correct. Final details were not arranged until May, but in the meantime Hill and Loveland negotiated a tentative agreement that they forwarded to Gould and Dillon for approval. Loveland still refused to lay the third rail between Denver and Golden, but he did agree to shift ore cars from narrow-to broad-gauge tracks at the expense of the railroad. This eventually proved so costly that the company added the additional track. Loveland also agreed to lay a third rail from his depot in Denver to the plant site so that coal, ores, and building materials arriving on other narrow-gauge lines could be shipped directly to the smelter. With the signing of this protocol the Denver and Rocky Mountain Railway passed into oblivion. Sullen over his defeat by Gould and Dillon, Loveland claimed that Hill's Road would have been little more than a feeder to the Colorado Central, but Gould and Dillon obviously failed to see it that way.[7]

Confident the railroad question would be resolved in their favor, Hill and his colleagues began searching for a new smelter location even before reaching an agreement with Loveland. In January 1878 Hill inspected potential sites on the outskirts of Denver. Then, after purchasing a plot on the corner of Fourteenth and Welton Streets, where he later built a spectacular house in the Second Empire style, he opened negotiations for an industrial tract just north of the city. The talks continued for a while, but when the owners refused to lower what Hill thought to be an exorbitant price, he broke off discussions. At this point the editors of the *Daily Times* warned that the high cost of real estate might persuade Hill to accept Loveland's suggestion to erect works in Golden. The danger was no doubt exaggerated, for Hill recognized Denver's superiority as a reduction center, but the fear prompted the Board of Trade to express its "disapprobation" of land speculators and its appreciation of the company's plans. Hill eventually resumed negotiations, and in April he spent $7,000 to acquire eighty acres of land just north of Denver, a tract bordered by the Colorado Central Railroad. In a classical mood befitting the Gilded Age, he and Pearce named the place Argo after the mythical vessel that carried Jason in search of the Golden Fleece.

Then came plans for the new smelter. Robert S. Roeschlaub, a noted Denver architect, designed the new plant with the assistance of Hill and Pearce. They intended to erect four major buildings—an ore house, a smelting plant, a coal house, and a refinery—as well as smaller workshops for carpenters, blacksmiths, and machinists, plus a two-story office tower at the front of the works. Spurs from the Colorado Central Railroad would enter the plant and connect the principal buildings and storage areas. Because of the fire hazard, Hill and Roeschlaub wanted to use brick as the primary building material, but as it was in short supply they switched to stone quarried at Morrison, a few miles south of Golden. They also selected corrugated iron for roofing. The plans called for the entire plant to be surrounded by a stone wall eight feet high. Outside the enclosure Roeschlaub designed a village with a hotel, church, school, and tenement houses. Argo would be a company town.

The new plant took shape during the spring and summer. Leaving Henry Wolcott in charge of operations at Black Hawk, Hill established temporary headquarters at Roeschlaub's office in Denver. In May he placed an advertisement in the *Daily Times* asking contractors to submit bids for brick and stone. As soon as Loveland's crews finished the third rail to Argo, the successful bidders shipped building materials by the carload to the construction site. By summer more than a hundred men were at work on the plant. When Hill gave city officials a tour in September, the major structures were in place, machinery lay on the grounds, and ores were piling up for reduction.[8]

The Argo smelter commenced operations on January 1, 1879, even though it was still unfinished. During the next few months Pearce and Wolcott pushed construction at a rapid pace, but work on the refinery went slowly because the enterprise took special precautions to preserve the mechanics of

Pearce's secret process. Not until the end of the year did the plant become fully operational. By then the firm even had its own steam engine to run between Argo and Denver and do switching at the works. During its first twelve months in operation the smelter's production of gold, silver, and copper rose to $2,450,000. This was a new high for the company, but it marked the last time Hill's output would be the largest in Colorado's reduction industry.[9]

The cost of relocation had been high. At the annual meeting in Boston in May 1879, James W. Converse, the firm's president, reported that debt had risen to $262,000, more than half the outstanding equity capital. To provide the smelter with additional working funds and reduce the huge financial obligations to a more acceptable level, the stockholders voted to double the capitalization from $500,000 to $1,000,000. They authorized the creation of 5,000 new shares of stock, which the enterprise sold within a year. The officers used part of the money to reduce liabilities. Hill took advantage of the opportunity to increase his holdings from 520 to 1,040 shares, displacing J. Warren Merrill, the corporate treasurer, as the largest stockholder.[10]

As Argo took shape on the north Denver plain, Hill took advantage of the firm's success to advance his political career. In May 1878 Jerome B. Chaffee announced that because of poor health he would not be a candidate for reelection to the United States Senate. No sooner had he made his decision public than it was rumored that Hill had decided to seek the seat. These rumors proved true. To spearhead his drive for the Republican nomination, Hill purchased a newspaper at Central City and filled the party coffers during the fall campaign. In the November election the Republicans won a majority of seats in both houses of the state legislature, thus making their nomination tantamount to election. A month later Hill emerged as the front-runner for Chaffee's seat, his drive headed by the Wolcott brothers, the new state senators from Clear Creek and Gilpin counties.[11]

As the leading Republican candidate, Hill became the target of vituperative attacks by the *Rocky Mountain News*, which was controlled by the future Democratic nominee, William A. H. Loveland. The paper sarcastically referred to Hill as the "Argo statesman" and accused him of numerous misdeeds and of spending the excessive sum of $15,000 during the fall campaign. The *News* also claimed that Hill had the support of the Atchison, Topeka, and Santa Fe Railroad, which wanted to control a senate seat in order to secure favorable legislation. When Hill unwisely attempted to purchase the paper in an effort to silence the criticism, the editors jeered that it could "be had at ten dollars a year, which is just as near as Mr. Hill will ever come to owning it."[12]

Despite the partisan attacks of the *News*, Hill's campaign for nomination and election went smoothly. When the Republicans caucused on the evening of January 9, 1879, Hill led the balloting from the outset and received the nomination on the fifth poll, besting his two nearest opponents, George M. Chilcott and Senator Chaffee, who at the last minute had decided to make the race. With Hill's election certain, the *News* could do little more than proclaim: "The Political Doom of Senator Chaffee Hermetically Sealed." When the general assembly convened several days later, Hill easily defeated Loveland. After both houses met in joint session to declare Hill the victor, he was formally introduced and made a short speech. In congratulating the senator-elect, the *Engineering and Mining Journal* commented that if he ran Colorado politics as well as he did his smelting enterprise, there would soon be "a moral dividend to declare, instead of the usual assessment."[13]

When Hill left to assume his duties in Washington, the day-to-day management of the Argo smelter fell to Richard Pearce and Henry Wolcott, and their leadership proved able indeed. They increased production from $2,500,000 in 1879 to $4,400,000 in 1884. The growth primarily reflected the rising output of mines situated along the South Clear Creek, in Summit County, and at Leadville, but the firm also purchased ores from markets in Utah, New Mexico, and Montana. South Park and Boulder County declined as sources of mineral, but small consignments shipped from Arizona, Nevada, and northern Mexico more than made up the difference. Although the mines of Colorado remained the smelter's chief supplier, the railroads linking the ore-producing regions of the West permitted the enterprise to serve a wide area, as Hill and his colleagues had foreseen.[14]

More than other mining regions outside Colorado, Montana remained essential to the firm's livelihood. Copper was the sine qua non in the Swansea process, but during the seventies it had grown scarcer in Hill's traditional markets on the forks of Clear Creek. Beginning in 1878, William A. Clark had shipped large quantities of copper ores from his mines in Butte to the furnaces at Black Hawk. Yet this arrangement had never proved satisfactory because transportation costs, amounting to forty dollars per ton, had eroded the profit margins of all concerned.[15]

Perceptive as they were, Hill and Clark recognized that smeltermen would soon erect plants in Butte. Copper production there was on the rise even though the Anaconda bonanza was still unknown. Hill and Clark wanted to steal a march on expected competition. Early in 1878, while the Argo smelter lay in the initial stages of construction, Hill sent Henry Williams, the manager of the Alma plant, to Montana to determine the feasibility of erecting works in Butte. Williams spent several weeks inspecting mines, locating building materials, and searching out sources of fuel and water. When he returned to Denver, he reported that prospects for a smelter were excellent.[16]

In June 1879 this report and the need of the Argo works for steady supplies of copper prompted Hill, Pearce, Henry Wolcott, Williams, and Clark to organize themselves into a body corporate as the Colorado and Montana Smelting Company. Six months later, after several unaccountable delays, they filed the articles of incorporation with the secretary of state in Denver and set the capitalization at $200,000, the same as the Boston and Colorado firm in 1867. Hill was elected president of the company even though he now held a seat in the United States Senate, and he purchased one-fourth of the two thousand shares issued. The remainder was not subscribed for another three years. Williams was appointed manager of the project and promised a fifth of the capital stock if he made the enterprise a success. Yet it was privately reported that he distrusted Clark's motives.[17]

With the company organized, Williams returned to Butte. He purchased the property of a defunct smelting firm, erected new furnaces along the banks of Silver Bow Creek, and began reducing Clark's ores to a matte averaging 50 percent copper and worth from $600 to $1,000 per ton depending upon the silver content. This he shipped to Argo for refining. As time passed, Williams increased the smelting capacity from eighteen to fifty tons a day and purchased ores from other mineowners. He reduced some consignments in Butte but shipped others directly to Colorado. When new companies erected smelters in Butte, he bought their matte for shipment to Argo. Although the two enterprises never merged into a single corporation, they remained closely associated—another example of the informal integration prevailing in the reduction industry.[18]

While operations went forward at Butte and Argo, the Boston and Colorado enterprise witnessed the appearance of a potentially dangerous rival. In 1881 a group of investors from Denver organized the Miner's Smelting and Reduction Company, which purchased and renovated a defunct smelter in Golden. Using a time-honored industrial subterfuge, the firm hired several of Hill's employees in order to obtain the details of Pearce's secret process and other techniques employed so successfully at Argo. The Miner's Company put its smelting furnaces into operation by summer, produced some matte, and extracted silver by the Ziervogel method. Then the metallurgists melted the gold-copper residue into a high-grade matte, the first step in Pearce's technique. Several weeks later Edward O'Niel, one of Hill's former employees, finished installing the equipment that duplicated the entire Argo method.[19]

Using this technology, the Miner's enterprise remained in business for another two years. It expanded operations at Golden, established sampling agencies in Black Hawk, Idaho Springs, and Georgetown, and lured more employees away from Hill's firm. Yet in spite of its technical success it failed in

1883, a victim of stiff competition and poor location. The *Engineering and Mining Journal* noted that the plant was situated in "an out-of-the-way place and could not compete with one more centrally located and with six times the capital," a lightly veiled reference to Hill's company, which once again had triumphed over a rival in the Clear Creek ore markets. To restore its monopoly on Pearce's process, the Argo firm purchased and razed the Miner's plant.[20]

Despite his continued success Hill saw the position of his enterprise change markedly. After more than a decade as the only viable smelter in Colorado, its production was surpassed in 1880 by the Grant works in Leadville. Even more significant, the swift development of silver-lead smelting on the North Arkansas thrust blast furnaces to the fore as the chief means of ore reduction in Colorado. Hill's company remained important for another three decades, but leadership in the industry had passed to the silver-lead entrepreneurs.

Hill and his associates had smelted ores at Argo for more than three years when fire destroyed the Grant works at Leadville in May 1882. This catastrophe threw the owners, Edward Eddy, William H. James, and James B. Grant, into consternation. Yet they were a resourceful group of men, and they had access to capital with which to rebuild. The question was where? Even before the conflagration, they had debated the possibility of relocating in Pueblo, at the foot of the Arkansas Valley. Here they would enjoy proximity to the coking coals of Trinidad and access to a good railroad system. Yet other considerations pushed Denver to the forefront in their thinking when the time came to rebuild the Leadville plant. Ever since he had gone into business in 1878, Grant had shipped his bullion to Omaha. In return, he had received preferential freight rates from the Union Pacific Railroad, whose president, Sidney Dillon, held a large block of stock in the Omaha Smelting and Refining Company. This was the key to the decision of Eddy, James, and Grant. They wanted to maintain their favorable relationship with the line and still lie on the most direct route from Leadville to Omaha. For this reason they decided to rebuild in Denver.[21]

After looking at several possible locations, Grant and his partners bought an industrial tract about two miles northeast of downtown and across the South Platte River from Argo. The land itself formed a terrace that sloped gently toward the waterway, and Grant and Malvern W. Iles took advantage of the contour when they designed the smelter. Their plans called for a system of railroad tracks laid in such a way that ores, fuel, and flux would be delivered directly to the upper levels opposite the charging doors of the furnaces. Then, in the course of operations, intermediate products would flow downhill to lower levels. Ultimately, smelter workers would run slag onto a dump and load bullion onto cars bound for Omaha.

Eddy, James, and Grant broke ground for the plant on July 2, 1882, and Denver's second great smelter took shape during the summer and fall. The partners purchased the best machinery and furnace equipment available from Fraser and Chalmers in Chicago, the Pacific Iron Works of Leadville, and the Colorado Iron Works of Denver. Grant hired crews of tradesmen who put up the brick and stone buildings, installed electric lights, and put in the most advanced technology of the day. Iles supervised all technical details. As work progressed, Eddy and James began shipping ores down from the mountains via the Denver South Park & Pacific Railroad. Huge stocks of fuel and mineral accumulated on the grounds as the smelter neared completion. After four months of steady work, Grant and Iles blew in the first reduction unit on October 7. Others followed in rapid succession, and by early November the plant was in full operation with eight blast furnaces, which had an ore capacity of 230 tons daily.[22]

Just as the smelter came on stream, however, Grant had to divert his energies from business to government; for, like that of Nathaniel P. Hill, his political star was ascendant. During the summer of 1882 the leaders of the Democratic party urged Grant to run for governor. His success in business had made him well known throughout the state, and party stalwarts wanted a strong gubernatorial candidate because they saw an excellent chance of winning in November. Grant was amenable to the idea. At the county conventions held during the late

Smelter worker drawing slag from a furnace at the Grant smelter in Denver. Harry H. Buckwalter, photographer. Colorado Historical Society, Denver.

summer, his supporters easily won majorities that led to his nomination as the Democratic standard bearer.

In sharp contrast to Grant's harmonious nomination, the Republican aspirants for governor created acrimony within the ranks of the party. The infighting at the county and state con-

ventions split the politicians into two factions—the Hill–
Wolcott group, known as the Argonauts because of their as-
sociation with the famous smelter, and the Jerome B.
Chaffee–Henry M. Teller group, known as the Windmills because of their alleged boastfulness. The Argonauts supported
Hill's right-hand man, Henry Wolcott, but the Windmills
blocked his drive and nominated their own candidate, Ernest
Campbell of Leadville. This worsened the disharmony between
the two camps, and the Argonauts refused to support
Campbell, thus setting the stage for a Democratic victory.
Neither Grant nor Campbell waged an energetic campaign
that fall; but when the votes were counted in November Grant
found himself elected the first Democratic governor in Colorado's history.[23]

Even after he took office, Grant, like other businessmen in
the Gilded Age, hardly let his public career interfere with his
entrepreneurial activities. This was all to the good, for even as
he assumed the governor's chair his firm was on the eve of
major changes. Early in 1883, Eddy, James, and Grant began
to explore the possibility of merging their enterprise with the
Omaha Smelting and Refining Company. What motives lay
behind this move remain conjectural, but the forces promoting
the talks must have included a desire to integrate operations
from sampling through smelting to refining, broaden ore markets, take advantage of the economies of scale, and raise capital. Talks ensued in the spring and summer, correspondence
flowed across the plains, and lawyers delved into legal matters.
Then negotiations culminated on July 5, 1883, with the creation of the Omaha and Grant Smelting and Refining Company.
This firm acquired all the property of its predecessors as well as
the sampling works of Eddy, James & Company, the "company" of course being James B. Grant.

The merger united some of the most important entrepreneurs in the industry—Eddy, James, and Grant of the Colorado firm, and Guy C. Barton, Edward W. Nash, Charles Balbach, and Joseph H. Millard of the Nebraska corporation. They
set the capitalization at $2,500,000 and specified in the corpo-
rate charter that the enterprise could not go into debt beyond
two-thirds of its equity capital and that the stock could not be

James B. Grant about 1882, when he was elected governor of Colorado. Colorado Historical Society, Denver.

increased to more than $5,000,000. Barton became president, Grant vice-president, and Nash secretary and treasurer. As a practical matter, Eddy, James, and Grant would handle all operations in Colorado.[24]

From the outset Barton, Grant, and their associates bent their energies to rationalizing operations and increasing the scope of their activities. First they rerouted ore shipments so that the Grant works in Denver received the bulk of the mineral under contract with shippers in Colorado. This lowered costs and permitted the Omaha smelter to concentrate on the ore markets of Utah, Montana, and Idaho. The firm also opened sampling agencies in Black Hawk, Aspen, and other camps to go with the plant at Leadville. And to further integration in

Denver, Grant and his colleagues built a sampling mill and a number of roasting furnaces with which to treat the rising production of sulfides.[25]

The firm also moved heavily into the Leadville ore market, which furnished about 60 percent of the mineral smelted at the Grant works. From 1882 through 1884 Eddy, James & Company bought ores whose value exceeded the production of every smelter remaining on the North Arkansas! Purchases became so extensive that they exceeded the capacity of the Denver smelter to process what the sampler had put under contract. To process the ores, the enterprise spent a reported $150,000 to acquire the plant of Cummings and Finn, who had erected one of the largest smelters in Leadville. Eddy, James, Grant, and one Henry Head incorporated the property as the Fryer Hill Smelting Company, although it remained a subsidiary of the Omaha and Grant Company.[26]

Operations in Leadville went well for a time, but in 1885 Grant and his colleagues found themselves beset with problems. Despite its size and strength the Omaha and Grant enterprise found itself unable to compete with the Pueblo Smelting and Refining Company for mineral high in lead. Rather than purchase ores at a loss, Grant and his colleagues withdrew from this phase of the market, but this meant they had to close the Fryer Hill plant. By the time the Pueblo smelter relinquished its hold on the market, Barton and Grant had enlarged both the Denver and the Omaha works. That made the Fryer Hill plant expendable, and it never reopened. Barton and Grant later leased the property to contractors who reworked the slag dumps to recover small amounts of silver and lead lost in the early days of processing.[27]

The closing of the Fryer Hill smelter hardly put a crimp in overall operations, but the experience still left a good measure of anxiety in the minds of Grant and his colleagues. With the production of dry ores increasing everywhere, they worried that future prosperity might be impaired if they failed to guarantee their plants a reliable source of lead. For this reason the Omaha and Grant Company bought the Terrible mine in Leadville for a reported $100,000, and metallurgists in Denver and Omaha were soon adding the concentrates to other ores "like yeast to the dough." The Terrible mine never provided enough lead to support the massive operations of the firm, but it provided some flexibility in a capricious ore market.[28]

By 1886 the Omaha and Grant Company had emerged as a highly integrated firm. It owned one mine outright, many sampling agencies, two reduction plants, a refinery, and a marketing arm well on its way to product differentiation with its brand, "Omaha lead." What was more, Eddy, James, and Grant, as well as their associates in Nebraska, owned large blocks of stock in many mining corporations, thus continuing the thread of informal integration seen throughout the smelting industry.

The success of Grant, Hill, and their associates drew the city of Denver to the attention of other entrepreneurs. One was Edward R. Holden, another his partner Richard Cline; still a third was Malvern W. Iles, and a fourth was Arthur Chanute. Each of these men had a different perspective on the reduction business. In Leadville Holden and Cline owned a prosperous ore purchasing agency through which passed large quantities of mineral destined for local works or the valley smelters. This gave Holden and Cline a broad perspective on the mining industry, just as it had Edward Eddy, William H. James, and August R. Meyer a few years before. Iles possessed the great metallurgical skills that had made so large a contribution to the prosperity of Grant's smelters and to the technological evolution of the industry. Yet, like Anton Eilers and Alfred W. Geist before him, Iles desired an entrepreneurial role. And Chanute was a prosperous Denver banker closely connected to the financial community in Omaha. Each of these men saw continued potential in the growth and prosperity of the silver-lead industry, and early in 1886 they organized the Holden Smelting Company.

Then came the questions of location and design. Denver was an ideal site. Both Iles and Chanute had business connections there, and the city now had even better railroad service than it had when Hill and Grant put up their plants. And so Denver was selected. Using capital probably raised by himself and

Chanute, Holden acquired about thirty acres of land near the Grant works. Then Iles, drawing upon his great experience and his knowledge of the latest advances in technology, designed an integrated smelter that was to be the most efficient of its day, a reputation it held well into the twentieth century. Since Iles had helped plan the Grant work, it was not surprising that his new structure resembled the older plant, although his design called for far more mechanization to reduce costs.

While Iles and Chanute did their work in Denver, the voluble Holden toured the mining camps in the high country, negotiating contracts for mineral. As might have been expected of an old ore buyer, he was energetic and successful. The first shipments of rock began arriving on the smelter grounds in May, several months before Iles planned to blow in the smelting furnaces. In the search for mineral, the mines of Leadville were particularly important to the firm—as they were to all the valley smelters except Hill's works at Argo— and the enterprise had the good fortune to enter the Leadville market just when the Pueblo Smelting and Refining Company lost its corner on high-grade galena. This later proved permanent, but Holden and his colleagues could not then be sure. To guarantee themselves adequate supplies of lead, they bought the Silent Friend mine in the Monarch district of the Arkansas Valley.

Meanwhile, Iles pushed construction along steadily during the summer and into the fall. The firm acquired good supplies of ore, iron flux, and El Moro coke, then in September Iles set the first smelting unit in blast. He added two more in October and a fourth in December. Excellent planning made operations go smoothly from the beginning, and by the end of the year the Holden smelter had shipped bullion worth nearly $533,000 in silver, gold, and lead, a creditable showing in so brief a period.[29]

For reasons that have always remained obscure, Holden and his associates waited until nearly six weeks after beginning operations to incorporate the new venture. Not until October 26 did Holden, Cline, Iles, and Chanute, as well as John L. McNeil and Samuel Adams, set their hands and seals on the articles of incorporation and fix the capitalization at $300,000. Of this amount they used about $200,000 to acquire property—meaning the reduction plant—and the remaining $100,000 went to provide working capital. They also elected Holden president of the corporation.[30]

With such a propitious start, Holden and his colleagues expanded and integrated their operations further in the course of 1887. Much of the actual work fell to Malvern W. Iles. He directed the construction of a mill for crushing ores and matte, built additional roasting units, and installed two more blast furnaces, giving the works a total of six, with an aggregate capacity of three hundred tons daily. Iles also experimented with a new process for treating zinc-bearing minerals excavated largely at Leadville, where miners were finding these ores in massive quantities as they blasted ever deeper into the earth in search of silver and lead. He had little success, but his efforts presaged the day when the industry would process huge amounts of zinc-bearing rock. The company also purchased another fifty acres of land where the Colorado Central Railboard built several miles of sidetrack to aid the movement of ore and fuel.[31]

These improvements augured well for future prosperity; yet in the midst of expansion there were significant shifts in management and ownership. As in any business enterprise, growth required substantial infusions of capital, and Holden and his colleagues borrowed much of what they needed from the Colorado National Bank. At the same time—and perhaps as a condition of the loans—two major figures in the banking industry, Dennis Sheedy and Charles B. Kountze, acquired large blocks of stock in the company. They saw Holden as a promoter and speculator, a reputation he also had in mining and smelting circles. Sometime in 1887, Holden sold a substantial quantity of stock to Meyer Guggenheim, a Leadville mineowner, and the two men later announced plans to build still another smelter in Denver. This only served to convince Sheedy, Kountze, and others that Holden could not be trusted, particularly since he was letting the potential competitor Guggenheim study the internal mechanics of operation. But, what was worse, Holden's management had now brought the company to the verge of bankruptcy. Kountze had Sheedy determine if anything could be done to save the enterprise and protect the loans of the bank. And Sheedy concluded that the only thing for them to do was to take over the management. As a result, at a special meeting the directors elected Sheedy president of the corporation and as a face-saving gesture relegated Holden to the position of general manager. Soon after, Holden sold his remaining interest and left the firm entirely to join Meyer Guggenheim.[32]

By his own admission, Sheedy "knew nothing of smelting or ores"; yet this stout, moustachioed man with a receding hairline had been very successful in business. Though born in Ireland in 1846, he had grown up in the United States, first in Massachusetts and later in Iowa. He had relatively little formal education, but he had risen through clerking, freighting, merchandizing, cattle raising, and banking to a position among the wealthy in Denver. He was a self-made man in the classic mold of nineteenth-century individualism. When he became president of the Holden Smelting Company, he "realized that to accomplish anything he must master the details of the industry." Right away he hired a tutor, and for the next three years he pursued his "incessant study," reading every book available to acquire a technical and practicable knowledge of the business. "Night after night" he studied, and day by day he managed.[33]

Once Sheedy took command, he and his colleagues reorganized the firm as the Globe Smelting and Refining Company. In the new managerial structure Chanute became vice-president, John M. Walker secretary and treasurer, and Iles superintendent of the works. The new directorate included Chanute, Walker, Herman and Charles B. Kountze, Thomas H. Woodelton, William B. Berger (Hill's son-in-law), and Sheedy himself. Iles lost his seat on the board, but he remained with the firm and proved essential to its success in the years ahead. Sheedy's group also increased the capitalization to $1,000,000 and began a new town known as Globeville for their workers. Later annexed by Denver, this community, inhabited increasingly by people of eastern European origin, became one of the most famous ethnic neighborhoods in the city. Polish, German, and Slovenian tongues resounded in the plants, for some workers spoke not a word of English. And Slavic churches later were built as the newer immigrants tried to re-create the customs and traditions of the old country. In more ways than one, the Globe smelter justified its name.[34]

Under Sheedy's leadership the enterprise sought out minerals along the eastern slope and in central Colorado. It invaded markets on the forks of Clear Creek and competed effectively with its rivals in Denver and Pueblo for the output of the Mattie, Lamartine, and other properties. The firm also drew heavily on Leadville, still the leading mining camp in the high country, an effort that put further pressure on the companies still in business there. Sheedy's firm became one of the first smelting enterprises to process mineral drawn from the Moyer shaft of the Iron Silver Mining Company, which had developed this new bonanza drift as older ore bodies gave out. Then as the Colorado Midland and Rio Grande railroads looped around the mountains from Leadville to Glenwood Springs and on to Aspen, the Globe Company tapped large quantities of mineral from Pitkin County, which went into its boom years in the late eighties because of cheaper transportation and access to the Leadville and valley smelters. The mines on Aspen Mountain and nearby peaks produced riches that even surpassed Leadville for a short time in the early nineties. The Globe plant was both a cause and a beneficiary of the boom.[35]

Sheedy and his colleagues enjoyed much success in the mining camps of Colorado, but opportunity—and necessity—drove them to faraway ore markets, something made possible only by rail transport. The charter of the old Holden firm had specified that the company would tap mines in New Mexico, Utah, and Idaho, whose districts grew in importance as time passed and the smelter expanded. Idaho proved particularly significant because the newly opened properties in the Coeur d'Alene country, notably the Bunker Hill & Sullivan and the Tiger-Poorman, offered large quantities of mineral relatively low in silver but high in lead. Because the Silent Friend mine in

Colorado failed to provide the plant with enough base metal, Sheedy and his colleagues sought the Idaho product to mix with dry ores coming down from Leadville, Breckenridge, and other camps.

For this very reason the Republic of Mexico emerged as an important source of mineral. The Mexican rail system had grown at a slower rate than its American counterpart, but it nonetheless carried increasing quantities of high-grade galena north to border towns like El Paso, Eagle Pass, and Laredo, Texas, there to be shipped farther north to the Globe smelter and its rivals. At first, about 1884, the ores came in a trickle—to use an incongruous metaphor—but by the early nineties Mexican mineral had grown so important to Sheedy and his company, and the whole industry, that he hired Henry Raup Wagner—a Yale graduate trying to avoid family pressures to pursue a career in law—to coordinate the firm's multifarious operations in Mexico. Wagner did such outstanding work that he advanced rapidly through the corporate ranks and later emerged as an important figure in the industry before turning from "bullion to books," as he wrote many years later.[36]

Having prospered for several years, Sheedy and his associates formulated plans to integrate operations even further by building a refinery. When they announced their intentions in 1890, the editors of the *Engineering and Mining Journal*, which kept its eye on all phases of the minerals industry, wondered if the enterprise could acquire enough bullion to make such an addition profitable. The Globe's furnaces would never produce enough bullion by themselves, many other refineries were in operation, and the number of smelters was declining. Sheedy and his colleagues delayed construction for a time, but ultimately they were not bothered by the doubts of other people. In 1891 they borrowed $100,000, and the next year they broke ground for the refinery. Once operations commenced, the firm supplemented its own bullion production by reaching out to garner the production of smelters as far away as Washington and Mexico.[37]

By 1893 Sheedy's enterprise had emerged as one of the major

forces in the smelting industry, the plant highly integrated, its operations farflung. With its sampling and crushing mills, roasting units, blast furnaces, and refinery, the operation in Denver was an efficient producer with costs among the lowest in the business. Sheedy's ore buyers crisscrossed the North American mining districts from southern Canada to central Mexico and from Colorado to California. And the abilities of Holden, Sheedy, and Iles in launching and managing the firm permitted the city of Denver to maintain its position as the most important smelting center in the industry.

Despite its rapid growth, the Globe enterprise never overtook its rival, the Grant smelter, as the largest producer in Colorado. Yet both evolved along the same path, and only a continuous process of expansion at the Grant works permitted the smelter and the city to maintain their claim to primacy in ore reduction.

Early in 1887, shortly after the Globe smelter came on stream, Eddy, James, and Grant began installing more roasting units, giving them a total of twenty-nine, with an aggregate capacity of three hundred tons of sulfides daily. The firm also enlarged six of its ten blast furnaces, increasing the smelting capacity to about four hundred tons daily. And it augmented the size of the dust chambers attached to the smelting units in order to capture larger quantities of metal volatilized in processing.[38]

Like Iles and his colleagues, Eddy, James, and Grant looked to new methods of reduction in hope of lowering smelting costs and augmenting ore supplies. During the summer of 1887 they explored the possibility of using petroleum drawn from Florence, Colorado, as an alternative source of energy in at least some of their operations. Experiments lasted several weeks, but the tests revealed that coal, coke, and charcoal were easier to obtain and more economical to use than "black gold." The next year Eddy, James, and Grant turned their attention to the possibility of reducing the zinc-bearing minerals that hampered the smelting of silver-lead ores. In conjunction with the Iron Silver Mining Company of Leadville, they erected an experimental mill in Denver to remove zinc from ores mined in

the Iron Silver tunnels. The technique employed was a modifi-
cation of the well-known Plattner process, a method that combined roasting, chlorination, and electrolysis to remove zinc
and leave a residue ready for smelting. Both companies invested substantial amounts of capital in the venture, but the
project proved financially unsound, and the mill had to be
abandoned.[39]

Eddy, James, and Grant continued to draw heavily on the
ore production of the Rocky Mountains. In gross tonnage Leadville remained their most important source of mineral. Even
though carbonates had virtually disappeared from the camp's
output, the smelter purchased huge quantities of sulfides from
properties like the Ibex, Adams, and Silver Cord Combination.
Like their rivals in the Globe Company, Eddy, James, and
Grant entered the booming Aspen market once the Colorado
Midland and Rio Grande railroads made it feasible to ship the
production across the mountains and down to Denver. Grant
even purchased bullion from the tiny works of the Aspen Mining and Smelting Company until the firm found it more profitable to sell rather than to process its own ores. In the early
1890s, as Cripple Creek, a new camp scarcely seventy-five
miles southwest of Denver, emerged as the greatest gold producer in the Rocky Mountains, Grant and his colleagues found
a new source of mineral for the furnaces in Denver. But this
material fell into the category of dry ores, which compelled the
enterprise to seek ever-larger quantities of lead-bearing ores
outside Colorado.[40]

Like their counterparts working for other companies, the ore
buyers of the Omaha and Grant enterprise reached aggres-
sively into distant markets. When the Bunker Hill & Sullivan
firm opened its great veins of silver-bearing lead in northern
Idaho, Grant and his colleagues secured much of the output
and had it shipped to Denver and Omaha, mixed with dry ores,
smelted to bullion, and refined to metal. Representatives of the
company also bought silver-lead minerals in Utah, Nevada,
California, and New Mexico, and on one occasion they even
purchased a small quantity of platinum rock mined in Wyo-
ming. Reaching beyond the United States, Grant and his as-

sociates sent agents into Mexico, where they negotiated contracts for high grades of galena to be shipped north and mixed with American production. Yet, because of the close relationship the firm maintained with the Bunker Hill & Sullivan and other shippers in Idaho, the Omaha and Denver smelters never became as dependent on Mexican minerals as some of their rivals.[41]

During these years the Omaha and Grant Company enjoyed a large measure of prosperity. From the time of its organization in 1883 through the end of 1890, profits flowed into the treasury, some to be reinvested in the business, some to be paid out in dividends. The Denver plant earned $2,095,000, the Omaha works another $1,614,000, for a total of $3,709,000. From the returns Grant and his colleagues reinvested $1,100,000 in the enterprise, used another $707,000 to pay interest on working capital, and paid out $1,900,000 in dividends. For comparison, the nominal capitalization of the firm was $2,500,000.[42]

In the early 1890s Grant and his associates decided on an ambitious program of expansion; but having taken this decision they found themselves needing an estimated $1,000,000 in new capital. They wanted to renovate both the Denver and the Omaha plants and erect an electrolytic copper refinery at the Nebraska works to separate copper matte evolved as a by-product in the smelting process. The key question facing Grant and his colleagues was this: Should they issue preferred stock or should they sell bonds? They debated the alternatives for some time, finally settling upon a foray into the debt markets.[43]

So large a bond issue, however, could be sold only through investment bankers in the East. After approaching a number of companies, Barton, Grant, and their colleagues agreed to let the underwriting to the firm of Clark, Dodge & Company, a prominent financial house in New York. This enterprise in turn approached Lee, Higginson & Company in Boston, and Blake, Boissevain & Company in Amsterdam. With the New Yorkers taking the lead, the general partners in each concern made a thorough investigation of the Omaha and Grant enter-

prise. They looked at profits and prospects, projected growth
rate, liabilities, values of plants and other properties, and the
articles of incorporation. The record impressed the bankers,
and they agreed to market the bonds with an annual interest
coupon of 6 percent. In return, the Omaha and Grant firm
placed a mortgage on its two plants to guarantee the securities,
which it expected to pay off in twenty years. The underwriting
syndicate worried about a weakness developing in the financial
markets of Europe, where they intended to place some of the
bonds, but by the end of 1891 Clark, Dodge and Company were
ready to sell the securities in New York, Boston, Amsterdam,
and London. The smelting enterprise received the money it
needed the next year.[44]

Grant and his colleagues then launched their new program.
First they had to reincorporate the enterprise, because the
existing charter prevented the company from embarking upon
certain avenues of expansion. On January 23, 1892, Eddy,
James, and Grant, along with Barton, Nash, and others, reorganized the venture as the Omaha and Grant Smelting Company. They kept the capitalization fixed at $2,500,000, but the
new charter permitted the enterprise to acquire coal mines,
stone quarries, and ore-producing properties and to sell stocks,
bonds, and other securities as well as to purchase those of other
firms. Then the firm allotted its new cash: $100,000 for the
copper refinery in Omaha, $400,000 for improvements to the
Grant works in Denver, and $500,000 for additional working
capital and the purchase of property. When completed, these
improvements enhanced the firm's position within the industry; yet this round of additions came on the eve of hard times
that, paradoxically, would open new opportunities for the company.[45]

As Denver's two silver-lead smelters prospered under the
leadership of Holden and Sheedy, Barton and Grant, the Boston and Colorado Smelting Company maintained its strong
position under the hand of old and new management. In 1885
Nathaniel P. Hill ran for reelection to the United States Senate
but lost his bid for a second term partly as a result of the
ongoing feud between the Argonauts and the Windmills. Bitter

in defeat, he decried the "notorious corruptionists of large experience" who had precipitated his downfall, but this was little more than a slap at the winner, Henry M. Teller. Nonetheless, Hill returned to Colorado to direct the fortunes of the smelting company and undertake other business ventures. Two years later Henry Wolcott resigned as manager of the Argo smelter to pursue his many interests in mining, at which he was very successful, and to continue his political career, at which he was very unsuccessful—he lost a second bid for the governorship. With Wolcott's departure after nearly two decades with the enterprise, Hill named Richard Pearce manager of the Argo works. Crawford Hill and Harold V. Pearce also entered the firm's service about this time. Both worked largely as assistants to their fathers, whom they eventually succeeded.[46]

New faces also appeared in the eastern management. During the eighties the presidency passed from James W. Converse, chief executive of the enterprise since its inception in 1867, to J. Warren Merrill, Joseph Sawyer, and finally Costello Converse, son of the original president. New individuals from the Boston area took seats on the board of directors. Yet Hill himself continued to be the largest stockholder, though he never held a high corporate office.

The background of the work force also changed during this time. In earlier years Hill had hired American-born workers to supplement a few men who had emigrated from Swansea, but now he and his managers began to hire people from other parts of Europe, a very large number coming from Scandinavia, particularly Sweden, in contrast to the southern and eastern Europeans going to work in the Globe and Grant smelters. Yet, as Hill remembered later, he was careful not to hire too many men from any one country lest this create a clannishness that might prove inimical to the interests of the company. Many workers continued to live in the tenements or small houses that the company built, and their children went to school across the river in Globeville with the children of smelter workers at the Grant and Globe plants.[47]

Despite these changes, Richard Pearce continued his long-standing efforts to increase the size and efficiency of the reverberatory furnaces. When Hill and his colleagues opened the plant in 1879, each smelting unit had an ore capacity of about twelve tons a day, the largest size then in use. But as the years passed these models proved inadequate because minerals

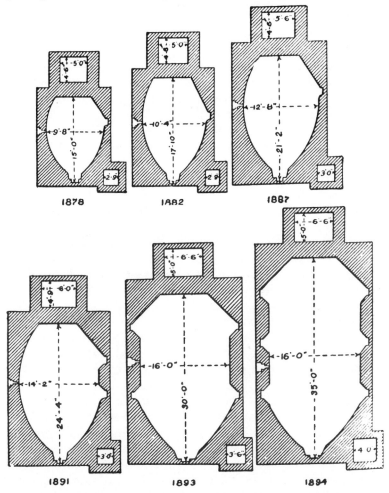

Fig. 5. Reverberatory development at Argo. By increasing the furnace dimensions, Richard Pearce enabled the smelter to reduce more ores at lower cost. Reproduced from Peters, *Modern Copper Smelting*, 7th ed.

declined in grade, silver-lead smelters competed for pyrites, and dry ores offered a new source if the firm could lower its costs. All this called for greater unit smelting capacity. Pearce responded by extending the length and width of the hearth and increasing the size of the firebox and smokestack. These modifications increased the daily capacity of each furnace to seventeen tons in 1882, twenty-four tons in 1887, thirty-five tons in 1891, and fifty tons in 1894. Pearce also began sending intermediate products directly from the roasters to the reverberatories instead of allowing the materials to cool for sampling and mixing, the normal practice in times past. Such economies of scale permitted the firm to lower its unit costs for fuel and labor and naturally translated into lower reduction costs per ton of ore.[48]

The new furnaces, however, presented Pearce with new challenges. The common practice for removing slag was to skim it from the surface of the molten matte and run it out through the slagging door. But the larger furnaces produced such enormous quantities of waste material that they created problems not seen in the past. If workers ran the molten slag into sand beds in front of the smelting units—as was the common practice—it made the building unendurably hot, wasted large amounts of space as it cooled, and proved expensive to carry out of the smelting house. To solve the problem Pearce fitted each furnace with slagging doors connected to iron troughs that carried the molten waste outside the building onto a dump. There it sizzled into a solid mass that Hill eventually sold to railroad companies for ballast on their lines.[49]

Pearce also evolved new methods for roasting ores before smelting. During the 1880s he replaced the old-fashioned calciners with Brown-Allen-O'Harra furnaces that had two hearths set one above the other. They eliminated the workmen who rabbled the ore charge, but despite the saving in labor the new roasters needed constant repair because hot sulfides damaged the moving parts. After enduring years of frustration and high maintenance costs, Pearce finally designed the prototype of what became known as the Pearce turret furnace. This was a significant improvement. Pearce and Crawford Hill subsequently organized a corporation to control the patent and sell units to silver-lead smelters, which also found the furnace useful. Later, Henry Williams and Richard Francis Pearce, the inventor's son, extended the design by adding tiers that reduced the consumption of fuel.[50]

The culmination of Pearce's technological success came in 1888 when he was elected president of the American Institute of Mining Engineers. As the topic of his presidential address, he chose the Argo process and provided the most articulate discussion of the method since Thomas Egleston's article more than a decade before; yet for "certain business considerations" he could not elaborate on the supposedly secret process. He also alluded to the failure of so many small smelters and the rise of a few large companies working in the urban centers of the West—a case of "the survival of the fittest."

Pearce's achievements in technology, combined with declining expenditures for labor and transportation, enabled Hill and his associates to lower their reduction costs, which meant they could pass on some of the savings to mining companies. (Competition from the silver-lead processors also forced them to do this.) As the eighties slipped by, the firm increased its returns to shippers, until by 1889 the enterprise bought ores for about 85 percent of their gross value, a remarkable rise from the early seventies, when Hill's "pay" hovered around 30 percent.[51]

During these years newspapers and trade journals occasionally mentioned rumors that Hill and his colleagues intended to build other plants or enter new aspects of the reduction industry. As early as 1883 the *Mining and Scientific Press* reported that businessmen from Tucson in the Arizona Territory had approached Hill about building a smelter in the sunbaked desert community. The editors were of the opinion that the firm could "afford to spare millions without any embarrassment whatever." Four years later, in November 1887, the editors of the *Engineering and Mining Journal* relayed speculation that the smelting company was about to erect blast furnaces and move into the silver-lead business, ostensibly because Hill, Pearce, and Wolcott had invested in certain mining properties. About a year after that, others claimed Hill was about to aban-

The Argo smelter about 1900. William H. Jackson, photographer. Colorado Historical Society, Denver.

Charging a furnace at the Argo smelter. Jesse D. Hale, photographer. Colorado Historical Society, Denver.

don his plant at Argo and relocate in Trinidad because the town had made an attractive offer that included very inexpensive supplies of coal. Finally, in 1889 another story intimated that Hill and his associates would erect a silver-lead smelter in Roaring Fork Valley. This plant would process low grades of ore unable to bear the cost of transport from the mines of Aspen to the furnaces of Leadville, Denver, and Pueblo. The men of Argo did none of these things, however, even if they considered such ventures seriously. Many years later Hill's brother-in-law, Jesse D. Hale, who held several managerial posts at Argo, wrote that management was very conservative and in particular did not wish to enter the highly competitive lead reduction industry.[52]

As the nineties dawned, Hill and Pearce continued to draw the bulk of their ore supplies from markets in Colorado. Mining companies on the forks of Clear Creek, the traditional source of mineral, sent carloads of pyrites via the Colorado Central Railroad down the narrow canyon to the Argo furnaces. And as the years passed Hill and Pearce competed effectively for the rising output of dry ores shipped from Leadville, Aspen, and other mining camps in the central Rockies. When Cripple Creek burst into its bonanza days as a fabulous producer of gold, Hill and Pearce found another lucrative source of mineral very close to Denver.

Yet, as they had done since the last years at Black Hawk in the 1870s, Hill and his colleagues still drew ores and matte from distant mining districts. They signed contracts with companies working veins and stopes in Idaho, Utah, Nevada, New Mexico, Arizona, and Mexico. Unlike their rivals in the silver-lead business, Hill and his associates never had to rely on the production of either Idaho or Mexico for the base metal essential in reduction. Yet this was small comfort. As the years passed, the Argo works grew ever more dependent upon the mines and smelters of Butte, Montana, for copper essential in what had become known as the Argo process. Hill and Pearce bought large quantities of both matte and silver-copper ores from the district, much of it coming from Williams's Colorado Smelting and Mining Company or the mines of William Andrews

The Globe smelter about 1900. William H. Jackson, photographer.
Colorado Historical Society, Denver.

Clark. Hill also purchased some supplies of matte from silver-
lead smelters who formed it as a by-product in reducing
copper-bearing ores.[53]

Even though Hill and Pearce searched far outside Colorado
for ores and matte, the value of metal shipped by the Argo
smelter reached a plateau during the eighties. After rising
steadily from the $271,000 worth of matte sent to Wales in
1868, the output touched $4,400,000 in 1884, a notable
achievement; but, after that, shipments from Clear Creek and
Gilpin counties fell off and the smelter's production slipped to
$3,700,000 by 1886. From this point Hill and Pearce increased
their output, but they failed to surpass the high of 1884 until
1890, when they marketed silver, gold, and copper worth
$5,000,000. Production then rose for two more years, reaching
an all-time peak of $6,060,000 in 1892.[54]

By the early nineties the names of Hill, Pearce, and Wolcott;

Eddy, James, and Grant; Holden, Sheedy, and Iles were familiar in mining camps from Canada to Mexico and from Colorado to California. They had built large plants, tapped distant ore markets, and carried the art and science of ore reduction to its most advanced degree. Yet, even as they transformed Denver into the great reduction center of the high plains, other smeltermen looked to Pueblo, a hot, dusty community that envisioned itself as "the Pittsburgh of the West."

Chapter 7

South by Southwest

THE CITY OF PUEBLO LAY ON THE SUNBAKED PLAINS OF SOUTHERN
Colorado, about 120 miles south of Denver, and like its major
rival it offered many advantages to industrialists. For one, it
sat only a scant 80 miles north of Trinidad, site of some of the
best metallurgical fuel in the United States. For another, the
old trading town lay at the foot of the Arkansas Valley, which
would allow ore cars from Leadville, Aspen, and other camps to
roll downhill to the smelting furnaces even more directly than
they could to Denver. And, for a third, Pueblo enjoyed good
railroad connections, the sine qua non for smelting mineral on
the Great Plains. The Denver & Rio Grande, the Santa Fe, and
numerous short lines all ran through town. By the late 1870s
Pueblo had taken its first step toward fulfilling its industrial
dreams when William Jackson Palmer and his associates in
the Rio Grande road formed the Pueblo Iron Company, a
forerunner of a large, integrated iron and steel plant.[1]

But the men who launched Pueblo's initial venture in the
silver-lead industry were wholly unconnected with railroads.
Alfred W. Geist and Joseph C. Mather were long experienced
in the mining and smelting of silver-lead ores, although rela-

tively little is known about either one. Geist, the son of a wealthy Bostonian, had studied for a time at the Sheffield School of Science at Yale University, but he was not inclined toward formal education. Before receiving a degree, he set out for Mexico, where he worked for a number of mining firms. Yet he did not remain there for long. The early seventies found him in Utah as a metallurgist for the Flag Staff Mining Company, which had built a smelting plant in the Salt Lake Valley. It was here that Geist must have made the acquaintance of Mather, who operated out of Salt Lake City as an ore buyer for the Omaha Smelting and Refining Company. Like Grant, Eilers, and others, both Mather and Geist aspired to be entrepreneurs themselves, rather than employees of other firms. When the Leadville bonanza made its first headlines in Utah, they decided to form a partnership, move to Colorado, and enter the reduction industry. No one knows whether it was by luck or by prescience, but in any event they selected Pueblo as the site of their smelter.[2]

Mather and Geist laid careful plans, for they were shrewd, experienced men entering what they knew was a highly competitive industry. First they acquired a building site on the south side of town near the junction of the Rio Grande and Santa Fe railroads; then they let contracts for building materials. Like some of their counterparts on the North Arkansas who were going into business about the same time, Mather and Geist had the wisdom to hire a skilled metallurgist to direct all technical operations. This man was August H. Raht, a graduate of the Bergakademie and a smelterman experienced in the Salt Lake Valley and elsewhere. He became superintendent of the works. Geist took charge of business affairs from the offices in Pueblo, while Mather toured Leadville and other mining camps to acquire adequate supplies of ore.

With the accolades of local businessmen, Mather and Geist broke ground for the new smelter in June 1878. They pushed construction vigorously during the long, hot Pueblo summer, and ores accumulated on the grounds as the plant neared completion. In early September Raht set the single smelting unit in blast. Processing went smoothly from the outset as the rich carbonate ores from the upper Arkansas, the San Juan, and other districts proved amenable to reduction. Mather and Geist were off to a propitious start.[3]

With ore production in Colorado rising sharply, Mather and Geist wasted little time in expanding the size and scope of their operations. Rumors appeared in October that they planned to triple the capacity of the works, and these reports proved true. Raht erected two more blast furnaces before the end of the year, and by January 1879 Mather and Geist had three units running full blast. They also installed a battery of roasting furnaces to aid the treatment of galena, pyrites, and tailings shipped from Black Hawk, Georgetown, and other camps. From the outset Mather and Geist were invading an ore market long dominated by the Boston and Colorado Company. They also purchased mineral in New Mexico, making their firm the first silver-lead smelter in Colorado to reach outside the state, a standard practice for the industry in the years ahead.[4]

As Mather and Geist prospered, they continued their efforts to expand and integrate operations. Like Hill, Grant, and others, they established purchasing agencies in the major mining camps in order to insure the furnaces at Pueblo of adequate supplies of ore and flux. They had Raht install more roasting units for the treatment of sulfides, and they appropriated enough capital for him to raise the number of blast furnaces from three to six, giving the plant a capacity of 240 tons of mineral daily. Having completed this work, Mather and Geist decided to carry integration one step further. In the summer of 1881 they had Raht build a refinery that made their plant the first major silver-lead smelter in Colorado to refine its own bullion. By the end of the year it was one of the most highly integrated in the industry.[5]

Mather and Geist had completed a remarkable program of expansion in little more than three years; but, as Hill and others had found before, rapid growth created problems of its own, although they were problems that any business would like to have. By the middle of 1881 Mather and Geist had come to realize that their enterprise had developed so rapidly that they could no longer operate efficiently as a partnership. Their chief problem was money—or, more accurately, a lack of money. They did not have enough working capital to carry on the business and still have adequate funds to make the improvements that were essential to maintaining their position in a highly competitive industry.

Mather and Geist saw only one solution. Sometime during the year they opened talks with local businessmen. Discussions went on all during the fall, culminating in December 1881 with the incorporation of a new enterprise, the Pueblo Smelting and Refining Company. Its directors included Mather and Geist as well as Charles Ballard, William Dwyer, and Robert McKenney. They gave the firm a nominal capitalization of $500,000, although it is doubtful that the five purchased the entire issue of stock.

With an infusion of fresh capital Mather and Geist and their new associates embarked upon another program of expansion and modernization. In April 1882 they broke ground for the construction of an auxiliary plant to manufacture lead pipes and shot, which they intended to market throughout the United States. When completed, the factory made the company unique in the silver-lead industry, although this step in forward integration followed naturally upon the building of a refinery the year before. Mather and Geist also used a portion of the new money to install electric lights throughout the works to aid night work and decrease the fire hazard, the bane of reduction companies. The firm also increased the tempo of its ore purchases, particularly in Leadville, the plant's most natural source of mineral. This effort portended difficult times for the smelters on the North Arkansas.[6]

Yet before the enterprise challenged the entire Leadville industry it went through another series of changes in management and ownership. The moving force in these events was Thomas Nickerson, a director and former president of the Santa Fe Railroad. He had long had an avid interest in southwestern minerals, which had always attracted New England capital. From time to time he and his associates had employed James D. Hague, Samuel F. Emmons, and other mining engineers to examine properties like the Breece Iron mine in Leadville. Nickerson wanted to enter the reduction industry, and in the early 1880s he considered erecting works in the Monarch mining district, which lay on a tributary of the Arkansas. He made a thorough investigation but decided against a venture in so isolated a region. Instead, Nickerson and his colleagues bided their time until late in 1882, when they quietly purchased control of the Pueblo Smelting and Refining Company.

Then came changes in management. Nickerson took over as president, while Mahlon D. Thatcher, a prominent banker in Pueblo, assumed the posts of secretary and treasurer. A. B. Laurie and Thomas Munn, both of Boston, accepted seats on the board of directors. Mather chose this time to leave the enterprise (in fact, he may well have sold out to Nickerson's group), but Geist remained as vice-president and general manager of the Pueblo works.[7]

Under Nickerson, Geist, and Thatcher, the Pueblo enterprise continued its steady program of expansion. Nickerson had ready access to the inner sanctums of investment banking houses in Boston, notably Kidder, Peabody & Company, and with this advantage the firm easily increased its capitalization to $750,000 in 1883 and to $1,000,000 in 1884. The new money enabled Geist to improve the efficiency of the plant, expand the lead manufacturing works, and broaden his sources of mineral. He also increased his penetration of the Leadville market, being fortuitously assisted by the roasting furnaces, which made it easier for his works to handle the rising sulfide production than it was for rival plants in Leadville, which had thrived almost entirely upon carbonates. It was partly as a result that in 1884 the Pueblo and Arkansas Valley companies signed a contract by which August Meyer and his lieutenants shipped large quantities of ore from their sampler down from the mountains to the Pueblo furnaces.

When the agreement with Meyer expired at the end of the year, Nickerson and his associates embarked upon their plan to gain control of Leadville's production of mineral high in lead. What sparked the challenge remains unknown. Perhaps

Nickerson and colleagues thought they could drive their high-country rivals out of business. Regardless, the Pueblo Company used its lower unit costs and large working capital to outbid its competitors for ores, and the Leadville smelters found they could not profitably compete. They had to purchase their supplies of base metal at a loss or blow out furnaces, and they did both. Even the prosperous Omaha and Grant enterprise withdrew from the ore market. Yet Nickerson's corner was short-lived. As the price of lead rose in the course of 1886, mining companies increased production and ended Nickerson's domination of those grades of mineral. More than anything else the rise in the output of lead-bearing ores saved much of the Leadville industry, at least for a time.[8]

Even though Nickerson and Geist had lost their corner on that aspect of the Leadville trade, they used their strategic position in the lower Arkansas Valley to entrench the company in many ore markets. They tapped the production of Aspen, Breckenridge, and other camps in the central Rockies. They bought mineral from districts on the forks of Clear Creek and in the high mountains of the San Juan. And they competed for rock in New Mexico, Arizona, and even more distant regions of the Southwest. Superior management, access to capital, and a close relationship with the Santa Fe Railroad added to the firm's competitive advantages. Yet the company still followed the same trends as its rivals in Denver and Leadville. As dry ores came into greater production, the firm had to dispatch its purchasing agents to Idaho and Mexico to get larger supplies of mineral with a high lead content.[9]

In the late 1880s the enterprise formulated plans to erect a copper plant to reduce certain classes of mineral mined in the Rockies. Since the firm could not generate internally the capital required, the directors decided to sell bonds bearing annual interest of 10 percent. Management sold the securities largely to the stockholders and erected the works, but as time passed the interest rate proved a heavy burden. In 1889, when the enterprise decided to expand the copper plant, the directors changed the debt and equity structure by fling a mortgage certificate with the Mercantile Loan & Trust Company of Boston and preparing new bonds carrying interest of 7 percent. These securities were offered to the shareholders at eighty-eight cents on the dollar along with two shares of common stock as an inducement to exchange the older notes. The offer proved successful. Management paid off the floating debt, expanded the copper plant, and bought two locomotives to do switching at the works. But, despite the thrust into the reduction of copper ores, the Pueblo Company still remained essentially a smelter of silver-lead mineral, its position secure in a competitive business.[10]

Pueblo also witnessed the emergence of a small, ethnically oriented neighborhood of smelter workers much like Argo and Globeville in Denver; but, in contrast to them, this one— known as Mexico—was not planned by the smelting companies. The community traced its origins to an earlier time when Pueblo was a trading post and Mexicans living there had built a cluster of small, tan adobe structures from the river valley up to the bluff on which Mather and Geist erected their smelter. Once the plant came on stream and Italian immigrants began taking over the lowest-paying jobs, they moved into Mexico in force and established a relatively self-sufficient community with markets, shops, and saloons. It was also said that the inhabitants cared more for Palermo and Sicily than they did for Pueblo.[11]

Despite their success—or perhaps because of it—Thomas Nickerson and his associates were not to have Pueblo all to themselves. By the early 1880s the Pueblo Iron Works had evolved into the Colorado Coal and Iron Company, which was fast becoming the only integrated iron and steel manufacturer west of the Mississippi River. And the town that styled itself the "Pittsburgh of the West" was soon to witness the construction of another silver-lead smelter, for so fine a reduction center could not escape the shrewd eye of Anton Eilers.

After dissolving his partnership with Gustav Billing on January 1, 1882, Eilers devoted his energy to mining endeavors for about a year. He made a long trip to Montana to investigate a property for a group of eastern investors and considered a journey to Mexico on behalf of another consortium. Yet he remained identified with ore reduction, and before long his reputation and skills as a metallurgist drew him back into the business.

Late in 1882 the owners of the Madonna mine, in the isolated Monarch district, hired him to solve a number of technical problems at their tiny charcoal-burning smelter. Eilers corrected the difficulties, but he told the Madonna owners that the works would never treat the ores profitably because they contained too little silver to overcome the high cost of processing. Yet Eilers added that the high percentage of lead could free a centrally located plant from the vagaries of production by mining companies. He suggested Pueblo as the best site for such a smelter.

Eilers's suggestions prompted serious talk. During the winter the owners of the Madonna mine, along with Eilers and some of his colleagues, agreed to launch an integrated mining and smelting firm that would take over the Monarch property, build a short-line railroad to join the main track of the Rio Grande, and erect a large reduction plant in Pueblo. These discussions bore fruit with the formation of the Colorado Smelting Company. Eilers was appointed general manager, although he was the leading spirit in the venture and later became president.[12]

With the company organized, Pueblo saw its second silver-lead smelter take shape in the course of 1883. For a plant site Eilers and his colleagues acquired an industrial tract south of the Arkansas River in what was actually South Pueblo. Eilers's design called for the construction of sampling works, roasting units, and four blast furnaces, although the enterprise erected only three units that year. Since Eilers intended to handle business affairs, he had to find someone to direct the metallurgical procedures. This was a serious question, but Eilers resolved the issue by luring his old friend Otto H. Hahn away from the American Mining and Smelting Company of Leadville. The firm broke ground in March, dirt flew in South Pueblo that summer, and a group of massive structures rose on the nearly barren plain. As the works neared completion, Eilers and Hahn conducted a series of experiments. They proved successful, and the Colorado Smelting Company began full-scale operations in September.[13]

After the plant had run steadily for several months, Eilers and his associates carried their original design to completion by erecting the fourth blast furnace, but they also expanded the smelter by installing other machinery. Then, with ever-larger quantities of sulfide ores coming to the works from Leadville and other camps, the enterprise appropriated additional money so that Eilers and Hahn could erect more roasting units. This work continued into 1884, when the price of lead declined and the Pueblo Smelting and Refining Company made its move in the Leadville ore market. Yet, with steady supplies of base metal coming from the output of the Madonna mine, Eilers and his colleagues avoided the competition. Their profit margins narrowed as the value of lead declined, but control of the Madonna mine saved the Colorado company from the ruinous struggle that developed on the North Arkansas.[14]

As the firm prospered, the plant became known throughout the industry as a training ground for metallurgists. The widely acknowledged preeminence of Eilers and Hahn attracted the attention of younger men wishing to begin their professional careers under the tutelage of the old masters. The competition for jobs—any kind of worthwhile job—was intense. One of the successful candidates was Arthur H. Dwight, a graduate of the Columbia School of Mines, who obtained his post through his uncle, Rossiter Raymond, who still maintained his close friendship with Eilers. Another was Eilers's own son, Karl Emrich Eilers, who took a position at the works about the same time as his college friend Dwight. Others came as well, and in future years they were known throughout the industry as "Eilers boys."[15]

It was well that Eilers was training new men, for metallurgists tended to be an itinerant group, never remaining long with one company. Change came in 1888 when Hahn resigned his position as superintendent after five years. His departure was a loss, but Eilers had a successor at hand in the affable Arthur H. Dwight. When he took Hahn's position, he also became the firm's chief negotiator with mining companies and local agencies that did sampling and "umpiring." A year later, after he had become more familiar with the job, he helped Eilers draw up plans to double the size of the smelter from four to eight blast furnaces. After the firm appropriated the requisite capital, Dwight and Eilers erected the new units in 1889.[16]

Eilers also used this time to expand the scope of his own interests in the smelting business. Many plants in the industry had grown ever more dependent upon ores mined in the Coeur d'Alene, but the costs of long rail shipments often proved burdensome, and both miners and smelters generally regarded freight tariffs as excessive, if not discriminatory. With the growing production of the Northwest served by only one major processor—at East Helena—Eilers and a group of New Yorkers controlling mining properties in Montana saw new opportunities for their capital and entrepreneurial talents.

Negotiations that ensued in 1887 culminated in the formation of the Montana Smelting Company. The founders of the enterprise included Abram S. Hewitt, well known as an iron manufacturer and mayor of New York City; Walter S. Gurnee, head of a large investment banking firm there; and Henry W. Child, an investor associated with the investment banking house of A. J. Seligman & Company, also of New York. They represented eastern interests, another manifestation of the control that region held over the industry. Yet there was a western component represented by Eilers and by Appleton H. Danforth, an obscure but important industrialist in Pueblo. The directors elected Hewitt president, but it was Eilers, the managing director, who designed the plant along the lines of the old Clausthal model and built it in 1888. Once in operation, this smelter tapped ores mined in the Northwest and increased the pressure on plants in Colorado in their search for base metal.

Yet the Montana Smelting Company was not to remain independent for long. Some time after Eilers had set the furnaces in blast, he and his associates entered negotiations with Samuel Hauser and others in the Helena and Livingston Smelting and Reduction Company, which had built the plant at East Helena, and with Barton Sewell and others in the

National Smelting and Refining Company, which had recently erected a refinery in Chicago. Like Eilers, both Hauser and Sewell had spent many years in the silver-lead industry, and now they merged their talents, interest, and firms into one enterprise, the United Smelting and Refining Company, chartered in 1890 under the laws of the state of New Jersey. This firm took over the plants at Great Falls, East Helena, and Chicago, and it acquired an option to purchase the Colorado Smelting Company. These facilities made the United enterprise one of the most highly integrated in the industry, one that compared favorably in size and strength with the Consolidated Kansas City and the Omaha and Grant firms.[17]

The major smelting enterprises at Denver and Pueblo enjoyed a large measure of prosperity during the 1880s. They had excellent management, access to capital, and superior locations—which is why they had chosen Denver and Pueblo in the first place. Yet not all the valley smelters were successful. Survival in the industry required far more than a favorable location.

Shortly after the Colorado Smelting Company started its furnaces in 1883, rumors appeared that another group of investors were considering the town as the site for a plant. The reports were substantiated in January 1884 by the publication of a letter written by Hiram Blaisdell, president of the recently organized New England and Colorado Mining and Smelting Company. He proclaimed that the enterprise would draw on Boston capital to erect a huge, integrated complex that would have an ore capacity of three hundred tons a day and would tap mining districts throughout the West.[18]

Blaisdell and his associates purchased an industrial tract from the Colorado Coal and Iron Company and commenced building in March. A few months later, as the plant neared completion, a heavy thunderstorm descended on Pueblo, and during the downpour a bolt of lightning struck the main smokestack. High winds blew the structure down. The misfortune set back the target date for operations, but the firm rebuilt the stack and announced November 1 as the day it would fire up its blast furnaces. Blaisdell and his associates, however, had neither the foresight nor enough working capital to acquire adequate supplies of ore and coke. When the firm blew in its two (not six) smelting units near the end of the year, they could do little more than conduct a trial campaign.

Undaunted by its initial setback, the New England and Colorado Company struggled on. Later that winter Blaisdell had several roasting units installed and made more energetic efforts to buy mineral. Yet his forays into the Rocky Mountain ore markets brought little success, for by May 1885 the firm had only enough mineral to keep one smelting unit in blast. This year also proved to be an unfavorable time to tap the Leadville trade, the smelter's most natural market, because the Pueblo Smelting and Refining Company chose this juncture to buy up ores high in lead. Like their rivals on the Great Plains and the North Arkansas, Blaisdell and his colleagues could not get adequate supplies of base metal. Operations were sporadic throughout 1885 and ceased entirely at the end of the year, when the enterprise failed. The plant then lay idle until it was sold at a sheriff's sale in 1889.[19]

During the early 1880s the large smelters at Denver and Pueblo grew into the dominant forces in the industry. The first rivals to feel the brunt of new competition were the tiny mountain plants. Almost simultaneously with the rise of Leadville, other mining towns had sprung up. Although none had the mineral resources to rival the carbonate camp, each flourished for a time, and all of them—places like Ten Mile, Kokomo, Gunnison, Silver Cliff, Breckenridge, St. Elmo, Gothic, Tin Cup, and Summit—witnessed a wave of smelter construction. Perhaps as many as three hundred different plants came into being, some owned by independent firms, others by mining companies. Yet in every case their costs were high, their volume low, their operations sporadic. As soon as railroads building through the Rockies reached—or even approached—them, mineowners signed contracts with the valley smelters, and this put the isolated local plants out of business. Hundreds of investors must have lost heavily, but the mining industry benefited because small mountain plants could never economically process the huge quantities of lower-grade minerals on which any mining company had to depend once the rich surface ores played out.

While forcing the small firms out of the industry, the valley smelters also undercut their major rivals in Leadville. Hard times for the companies there began about 1883, spurred in part by competition, lower grades of ore, and the steady appearance of sulfides. Yet the Leadville works held such a location that they could also tap ore markets in Breckenridge, Kokomo, Aspen, and other towns. This position helped them remain in business, although they were under constant pressure. The valley smelters also undermined the four small smelting companies that had located at Golden in the late seventies, forming what was known as "smelters row." Except for the tiny Golden Smelting Company, which struggled on until 1888, these firms had failed by the early years of the decade.[20]

The founding of the Argo and Grant smelters, coupled with the decline of Leadville, established Denver as the greatest ore reduction center in Colorado, if not the nation. But this supremacy did not go unchallenged, for Pueblo remained an ardent rival. In both the Colorado Smelting Company and the Pueblo Smelting and Refining Company, the parched community on the lower Arkansas had two enterprises that were great in their own right. And in the Colorado Coal and Iron Company it had the only integrated steelmaker west of St. Louis. But after Edward R. Holden erected what became the Globe smelter, Pueblo wanted another reduction firm to keep pace with its rival to the north—or to stay one step ahead, if you wished to count the Coal and Iron Company (some did). In any case, this aspiration brought the town fathers and the business community into contact not only with the ubiquitous Holden, but also with his new partner, Meyer Guggenheim, a short man with large, bushy sideburns.

Of all the great industrial capitalists of the late nineteenth century, Meyer Guggenheim almost certainly remains the most obscure. Perhaps he wanted it this way, for he seems to have craved anonymity. Yet he could not entirely avoid the written word. He was born in the Jewish ghetto of Lengnau,

Switzerland, in 1828, became a tailor in his youth, and practiced that trade in different parts of Europe. At the age of twenty he emigrated to the United States and settled in Philadelphia, where he did well in numerous business ventures. By 1875, when he was forty-seven years old, he had accumulated a fortune estimated at $250,000, a substantial sum for that time. He also had seven sons whom he expected to join him in business.

As he prospered in his adopted country, Guggenheim made his way into manufacturing and merchandizing on a large scale. He was, for example, a cofounder of the American Lye Company, a remunerative venture based in Philadelphia. Yet it was as a producer, importer, and distributor of lace fabrics and other goods that Guggenheim found even greater success. After selling his interest in the lye enterprise, reportedly for $100,000, he formed a partnership with Morris H. Pulaski, an old colleague in the spice trade. The two men then established manufactories at Saint Gall, Switzerland, and Plauen, Saxony, and shipped the products to New York and Philadelphia for sale in the United States. Guggenheim's sons Daniel, Solomon, and Morris (known also as Murry) handled affairs in Europe, while Isaac, a fourth son, directed operations in New York. The firm of Guggenheim and Pulaski returned large profits, but by the early eighties both men perceived the superiority of New York to Philadelphia as the center of the firm's activities in the United States. This led to changes. In 1882 Guggenheim bought out Pulaski and reorganized the enterprise to bring in his four eldest sons as partners.[21]

The new venture was known as M. Guggenheim's Sons. Although the name seemed formal and ponderous, it scarcely impeded progress in business, and most people simply referred to the enterprise as "the Guggenheims." The firm represented a partnership between Meyer, who had invested about $300,000, and Isaac, Daniel, Morris, and Solomon, who had advanced another $137,000. Without delay the Guggenheims moved their operations to New York City and opened their new offices at 350 Broadway on October 1, 1882. The venture grew rapidly, and by the middle years of the decade the annual sales amounted to more than $2,000,000. The Guggenheims established a reputation as capable businessmen, and their credit stood high in the financial circles of New York.[22]

As Meyer Guggenheim prospered as an importer of European goods, the specter of silver lured him into western mining, as it did many others. In 1879 Charles H. Graham, a friend in Philadelphia, traveled west to Leadville, then riding the crest of its early boom. He formed a partnership with George Work, Samuel Harsh, and Thomas Weir, from Philadelphia. The four speculated in mining properties, and they happened to open negotiations with an old prospector. This was A. Y. Corman, who owned two mining claims in California Gulch—the A. Y., named for himself, and the Minnie, named for his wife. Like most prospectors, Corman was eager to sell. After a round of talks Graham and his associates bought the properties for $4,000. As the story goes—and it may not be accurate—Harsh was the only one of the partners with money enough to pay. He lent his friends the funds they needed, but when the notes came due Graham, Work, and Weir could not repay. Rather than forfeit his share in the mines, Graham approached his friend Guggenheim, who agreed to reimburse Harsh, on one condition. The partners would have to sell him an interest—a quarter share according to some accounts, control according to others. This satisfied Graham and his associates—it had to—and so for a reported $5,000 Meyer Guggenheim became part-owner of two partially developed mining claims on Iron Hill.

Officially, the firm of Graham & Company took control and began the development of the A. Y. and Minnie mines. Harsh became general manager, but it was the wealthy Guggenheim, residing far away in Philadelphia, who provided most, if not all, the capital to purchase machinery, sink shafts, and pay the hired hands. Harsh pushed the work cautiously as the winter of 1879 rolled on into 1880, and an anxious Meyer Guggenheim waited in Philadelphia for good news. Time passed, and several months later Harsh telegraphed that he had encountered an ore body holding 60 percent lead and fifteen ounces of silver per ton. This was enough to insure the enterprise of profitable operations, although it was hardly a bonanza. Development continued, but not until July did Harsh send his first ores to the smelters in Leadville. Over the next two years he shipped about fifty tons daily, a respectable figure, but one that did not compare with the output of the Iron Silver, Robert E. Lee, Morning Star, and Silver Cord Combination.[23]

As the years passed, Guggenheim and his associates saw their operations frustrated by changes in the ore bodies. As happened sooner or later in Leadville, the easily processed carbonates produced by the A. Y. and Minnie gave way to sulfides, but they were richer than the oxidized minerals above—a very unusual phenomenon. Yet they also held large amounts of zinc, which created new problems because the smelters penalized such ores heavily. Deductions ranged from fifteen to twenty-five dollars per ton. Guggenheim and his colleagues erected a mill to remove the zinc, but the process failed to prove out. Harsh pressed forward with development in hope of locating more carbonates, but all he encountered in the drifts and stopes were huge quantities of low-grade zinc sulfides. In 1889 the firm commissioned three well-known mining engineers, Walter Renton Ingalls, Philip Argall, and Henry E. Wood, to study the situation. They concluded that the A. Y. and Minnie held at least 500,000 tons of ore worth about fifteen dollars each in terms of silver and lead, but the smelter penalties on zinc made the mineral worthless.[24]

Partly because of zinc and partly because he disliked sharing profits with smelters, Meyer Guggenheim developed a growing interest in the ore-reduction industry. He had absolutely no background in metallurgy, but then neither did Thomas Nickerson, Dennis Sheedy, and others. About this time Guggenheim began to discuss the possibility of erecting works with Edward R. Holden, the old Leadville ore buyer, now a leading figure in the smelting industry. Late in 1887, when Holden needed cash, he sold Guggenheim a large block of shares in the Holden Smelting Company. At once Meyer began a meticulous study of operations and began to formulate plans for a venture of his own.[25]

Before undertaking such a project, however, Meyer had to

resolve a double problem. He wanted his three youngest children, Benjamin, Simon, and William, to come into the firm of M. Guggenheim's Sons as equal partners, and he wanted this enterprise to erect the smelter through a subsidiary corporation. But as the story goes, the four eldest sons, Isaac, Daniel, Solomon, and Morris, opposed both ideas, and with good reason. The younger brothers had not contributed to the success of the import business, nor did anyone in the family know much about the workings of the highly competitive smelting industry. Yet Meyer was determined to have his way. While he negotiated with Holden, he won over his four oldest sons, at least in part by promising to underwrite any losses in the reduction venture with profits from the A. Y. and Minnie mines.[26]

With the problem resolved, Meyer Guggenheim pressed forward with his plans to engage in the smelting business. On January 13, 1888, Holden and his associate Richard Cline, together with Meyer, Daniel, and Benjamin Guggenheim, organized themselves into a body corporate as the Denver Smelting and Refining Company. They set the capitalization at $500,000, divided into five thousand shares of stock. Meyer took 51 percent of the issue for M. Guggenheim's Sons, thus assuring his family of control. Holden and Cline, and perhaps Graham and Harsh, purchased some of the remainder. What was left went into the corporate treasury.[27]

Once organized, the new enterprise burst into the headlines. Holden, whose public profile was as obvious as Meyer's was inconspicuous, told a newspaper reporter that the firm intended to construct the "most complete establishment ever built in the United States." The plant would have six blast furnaces with a capacity of four hundred tons daily and a refinery as well. The cost would be $200,000. After this brave announcement, Holden and Guggenheim opened an office at 1657 Arapahoe Street in Denver. They hoped to build here because of the superior railroad system, but they also were considering Leadville, Colorado Springs, and Pueblo.[28]

Despite their desire to build in Denver, Holden and Guggenheim had trouble finding a suitable site, and this gave

Pueblo an opportunity to make a strong pitch. Seeing a chance to obtain a third smelter, the Board of Trade asked Holden to come down to southern Colorado for a talk. He accepted and went there in March 1888, accompanied by Benjamin Guggenheim. Then the members of the board presented their case. Leadville, they said, suffered from high labor costs, problems in ore supply, and discriminatory freight rates. Denver was too far away from coal, lime, and coke. Colorado Springs had nothing to offer. But Pueblo, claimed the board, had an excellent transportation system. It was near the coking coals of Trinidad. It was closer to Mexico than other smelting centers in Colorado. And it had a plentiful labor supply that would make construction and operation cheaper than in Denver or Leadville. After listening to the arguments—all persuasive, Holden and Guggenheim returned to Denver. There a reporter asked if they had decided on a site. "Not positively," replied Holden, "but it will not be Denver."

Events now gathered speed. The partners discussed the matter as best they could, then Holden and Benjamin Guggenheim made plans to return to Pueblo for another round of talks with the town fathers. The second trip came in late March. By this time Holden and the Guggenheims—mostly the latter—had probably decided to build the smelter in Pueblo, but they concealed the decision in order to gain concessions from the city. And Pueblo's first citizens left nothing to chance. As their train pulled into town, Holden and Guggenheim found themselves greeted by two bands, the mayor, bankers, county commissioners, real estate promoters, and a group of other dignitaries. After the customary handshakes, the group formed itself into a parade that wound slowly up the streets of the city to the offices of the Board of Trade.

Once the hoopla died away, the smiles and glad hands of the railroad station melted into sober thought as Holden presented his proposition. First, he wanted Appleton H. Danforth to renew his offer of land outside the city. Danforth agreed, since the property was nearly worthless, not even useful for grazing. Then came the second proposal. Holden enumerated certain tax advantages that he wanted the town to grant the enterprise. The members of the city council agreed to arrange matters. Then came the third proposition. Holden wanted the Board of Trade to advance a bonus of $25,000 in cash, although he announced as a fillip that the firm would break ground within a week if the money were forthcoming along with the other considerations.

Danforth's gift of land and the council's concessions on taxes were one thing, but $25,000 in cash was another matter. Pueblo wanted a third smelter, however, and so the Board of Trade decided upon a gala town meeting to open a fund-raising campaign. On the night of March 29, the town's elite gathered at the opera house for a festive evening. Band music and eloquent speeches punctuated the spectacular. Local singers sang a version of "Marching through Georgia," in which "Holden" rhymed with "golden." Then the promoters introduced Edward R. Holden himself, the crowd cheered, and the promoter made a few modest remarks. Now for the $25,000. The members of the Board of Trade called for contributions. The mayor volunteered $100. Governor Alva Adams, a Pueblo man, proffered $500. Civic pride was at stake, more money flowed in, and the board organized groups to canvass the town for additional donations. Then, in keeping with the festive spirit of the night, the city council voted to waive local taxes on the company's property for ten years. Holden was lionized, but scarcely anyone noticed Benjamin Guggenheim, and Meyer Guggenheim was not even there.

Two weeks later, on Tuesday the tenth of April, Holden and Benjamin Guggenheim again returned to Pueblo, this time to hold a press conference. They announced that the enterprise had completed its plans for a smelter to be built outside the city limits. Title to the land was secure, the tax question settled, and $25,000 safe in the company's coffers. They hoped to have the works in operation by the first of October.

Then came construction. Holden and the Guggenheims engaged E. P. Kirby, an experienced engineer, as superintendent of the works. He directed much of the actual building—or at least he tried to. As construction went forward during the summer heat, the Guggenheims were everywhere, suggesting alternative approaches they thought would save both building and operating costs. This of course was sound business practice, but neither Meyer nor his sons knew much about this particular business. Meyer also wanted to avoid using patented devices like the siphon-tap of Arents and Keyes so the enterprise would not have to pay royalties. Tempers flared. Years later one of the disgruntled builders wrote that Meyer and Benjamin were "always wanting to change the plans or propose innovations that metallurgically would have been ridiculous. The greatest difficulty . . . was pester."[29]

Despite the growing animosity among the builders, the plant neared completion during the fall. By September Kirby had just about completed the ore bins, sampler, and roasting house, but machinery failed to arrive on schedule, and the target date for beginning operations had to be moved forward from October to December. Yet ores from Aspen, Leadville, and other camps accumulated on the smelter grounds, particularly large consignments coming from the A. Y. and Minnie mines, which Meyer controlled. The partners also changed the corporate name to the Philadelphia Smelting and Refining Company in honor of the Guggenheims' first home in the United States.[30]

Holden and his colleagues commenced operations in December, but no sooner had the smelter gotten under way than serious problems developed. For one thing, the blast furnaces "froze up" when workmen tapped bullion in the course of reduction. This occurred because Meyer had insisted upon using a hearth and stack arrangement for withdrawing bullion rather than paying Arents and Keyes for the right to use the siphon-tap. For another thing, the walls in the engine house cracked and the floor sagged badly. And a third problem involved ores—they were so refractory that no other smelter would purchase them, said one observer. Benjamin Guggenheim disclaimed all such allegations, but after a few weeks no one could deny that operations were going badly. To make matters worse, the price of silver declined, Kirby quit, and appalling losses grew on the books. Holden and Guggenheim replaced Kirby with Herman Keller of the Globe smelter, but in the short run he could do nothing about the large doses of red ink. One source estimated the losses at $50,000 a month.[31]

As the continuing recital of bad news arrived in the Guggenheim offices in New York, Meyer finally had to sit down with his seven sons to discuss what should be done. It was a gloomy meeting, not without recriminations about the wisdom of entering the industry. Faulty metallurgical procedures, falling silver prices, managerial quarrels, and hard-to-process ores now threatened the family fortune built up so assiduously over the years. Yet Meyer remained hopeful. He and his sons finally decided to increase the firm's capitalization from $500,000 to $1,250,000 and use the new money to remodel the furnaces, purchase better grades of mineral, and hire an expert metallurgist who could resolve the technological impasse.[32]

Meyer Guggenheim had the financial wherewithal to withstand such heavy losses—at least in the short run—but his associate Edward R. Holden did not. And neither did Richard Cline. They wanted to sell out before they lost everything, and Meyer agreed to purchase their interest on behalf of the firm of M. Guggenheim's Sons. Yet the price was heavy. To raise the money, Meyer had to sell his own stockholdings in the profitable Globe enterprise to Dennis Sheedy and Charles B. Kountze.

Having acquired new capital and parted with Holden, the Guggenheims had to find a competent metallurgist who could resolve the technological problems at the smelter. First-rate technical experts still drew premium salaries in the industry, as they always had, but having lost heavily because of Meyer's own penury, the Guggenheims now paid top rates to obtain the services of August Raht, long known in the business as a "trouble man." Raht had been instrumental in the rapid growth of the old firm of Mather and Geist, and after leaving them he had erected smelters in New Mexico, Utah, and Montana. Now this itinerant engineer returned to Pueblo as metallurgist of the Philadelphia smelter. Using some of the new capital, he remodeled the blast furnaces and installed other machinery essential in the most efficient reduction of mineral. Eventually Raht had the plant operating in proper fashion.

Yet Raht's work, essential as it was, represented only one

change transforming the Philadelphia smelter into a profitable venture. Simon Guggenheim crisscrossed the high country in search of better grades of ore, and other purchasing agents in the Sierra Mojada negotiated what was reported to be the largest contract ever signed between a Colorado smelting firm and a Mexican mining company. Benjamin Guggenheim, meanwhile, continued as general manager of the Philadelphia works, much wiser and more knowledgeable than he had been the year before. But Herman Keller quarreled with William Guggenheim, Meyer's youngest son, who did not get along even with his brothers. Keller quit as superintendent and left the company.

Like their rivals in Denver and Pueblo, the Guggenheims and their managers hired many recent immigrants, mostly for lower-paying jobs. Most of these people had come from Italy and Austria-Hungary, and many lived in the boardinghouses or small homes built near the plant. Other workers, probably those with better positions, lived farther away, where they bought or rented more substantial dwellings. Yet, as elsewhere, the smelter workers' life was not easy. They worked a twelve-hour shift for pay ranging from $1.75 to $4.00 a day, the standard in Denver and Pueblo. The heat of the furnaces, coupled with the higher temperatures of the southern plains, also increased labor turnover as some workers quit in the summer to find other jobs or prospect for mines in the cooler mountain climate.

The Guggenheims were not immune to the growing labor unrest of the day, and in September 1889 they sustained a setback in their efforts to revitalize the Philadelphia works. The Guggenheims had granted their workers an eight-hour day during the hot summer months in Pueblo, but in September, when the family tried to reinstitute the normal twelve-hour shift worked throughout the industry, some workmen went out on strike. Like other businessmen in the Gilded Age the Guggenheims blamed the walkout on a union that intimidated most of the employees. After a short time the strike failed and the twelve-hour day returned.

Once the Philadelphia works started making a profit, the

Guggenheims established themselves as a major force in the industry. Simon Guggenheim and a new associate, Edgar L. Newhouse, proved themselves able competitors in the ore markets of Aspen, Leadville, and elsewhere. Benjamin Guggenheim emerged as an effective plant manager, and August Raht provided the metallurgical knowledge that enabled the smelter to process mineral shipped from mining districts throughout the West. By the early nineties the Guggenheims were on their way to preeminence in a business to which they were truly latecomers, although their late entry would prove to be a distinct advantage.[33]

As Denver and Pueblo emerged as great centers of ore reduction, the smelting industry developed in a somewhat parallel way in the distant, spectacular mountains of southwestern Colorado. Mining in the San Juan had gone on since Spanish times, but little systematic exploitation took place until the American era. Hostile Indians, severe weather, and the high, rugged peaks hindered work until the 1870s. Yet work there was, and with the location of silver-bearing lead ores, camps like Silverton, Lake City, and Ouray sprang to life. Smelting firms soon dotted the high country, but as elsewhere they were tiny, high-cost operations with indifferent success. They could foster but not sustain the development of a major mining industry. As early as 1878, at least one mineowner claimed it was cheaper to ship his production all the way to Mather and Geist at Pueblo than to send it to local works in the San Juan.[34]

The arrival of the railroad brought more prosperous times to southwestern Colorado. The Denver & Rio Grande Railway ran its first trains to Durango in 1880 and made this town, more or less south of the mining region, what Denver was to the Front Range and central Rockies, albeit on a smaller scale. Once it reached Durango, the road extended its tracks to Salt Lake City. About the same time, narrow-gauge lines, particularly those of Otto Mears, "the pathfinder of the San Juan," built north into the major mining camps. Like Denver and Pueblo, Durango emerged as a potential site for the smelting industry.[35]

As the Rio Grande threaded its track into the San Juan, the

idea of launching a reduction company appealed to William Jackson Palmer, the tall, distinguished-looking president of the line, and William A. Bell, a British financier associated with many of Palmer's ventures—most notably the railway and the Colorado Coal and Iron Company. Palmer, Bell, and their colleagues already owned mines in the mountains and extensive deposits of coking coal near Durango. With such diverse yet interrelated investments they perceived that a silver-lead smelter would integrate some of their interests and provide additional revenues for the railroad, the heart of their industrial empire. In the summer of 1880 this thinking prompted them to organize the San Juan and New York Mining and Smelting Company and acquire an existing smelter to serve as the nucleus of an integrated corporation.[36]

The plant acquired was the Silverton Smelting Works, a small operation erected by the firm of George Green & Company in 1874. The enterprise had only a few roasters and a single blast furnace with the tiny ore capacity of ten tons daily. Although Green had a large personal fortune stemming from investments in Iowa, the smelter had worked only sporadically because of the long winters, high costs, and abysmal transportation system of the San Juan. Yet the works had an excellent metallurgist in the mercurial John A. Porter, a Freiberg graduate who had been an early member of the Eureka group and who may have pioneered at Silverton the use of water jackets to cool silver-lead furnaces—an assertion long in dispute. In any event, Palmer retained Porter as superintendent and let him buy into the San Juan and New York enterprise.

Planning now entered a more advanced stage. The new venture acquired several mines outside Silverton, deposits of coal in the Animas Valley, and a limestone quarry near Durango, for Palmer and his associates wanted the enterprise to own everything required for reduction except ores, most of which would come from mines across the San Juan. Then construction began in Durango. As Porter smelted into bullion what ores remained on hand in Silverton, he and his associates erected their new plant. They had the works ready for operation by April 1881 and blew in their two blast furnaces on the sixteenth of the month. Porter later converted the Silverton smelter into a sampling agency to obtain ores for the Durango plant.[37]

Despite its strategic location, the San Juan and New York smelter never had the same prosperity as its rivals on the high plains. Porter saw his reduction costs fall below what they had been at Silverton, but they never declined to the level of the plants in Denver and Pueblo. Companies there competed effectively in the San Juan ore markets. And to the undoubted disappointment of the stockholders, Porter's operations continued to be sporadic because the severe winters curtailed mineral production in the San Juan and hampered the steady shipment of ores, fuel, and flux. Porter had to keep his furnaces out of blast for weeks at a time, and production remained relatively small throughout the course of the decade. Nonetheless, to remain competitive, the enterprise remodeled and expanded its smelter. In 1885 Porter enlarged his blast furnaces, nearly doubling the capacity of the plant. He also installed more roasting units to process sulfides. Yet even with these improvements the smelter remained marginal.[38]

During this time the relative isolation and slow development of transportation systems within the San Juan permitted a group of small mountain smelters to remain in business there far longer than their counterparts in the central Rockies. At Lake City the firm of Crooke Brothers and its successor, the Crooke Mining and Smelting Company, Ltd., processed ores until the mid-eighties, when stock manipulation and financial chicanery drove the corporation from the industry. At Rico both the Grand View Mining and Smelting Company and the Pasadena Mining and Smelting Company began operations in 1880 and continued working until late in the decade, when decreases in transportation and reduction costs allowed the Durango and Pueblo smelters to drive both firms out of business.[39]

Even as these firms failed, the New York and San Juan enterprise lay on the eve of major changes. Despite the improvements financed by Palmer and his associates, the venture still operated at a loss. This was hardly satisfactory. In the winter of 1887 Palmer and his colleagues discussed what it would take to convert the smelter into a profitable venture. The talks involved the major stock- and bondholders as well as investment bankers at Henry Amy & Company, and at Spencer Trask & Company, which had always worked closely with Palmer's group. In the end it was decided to reorganize the firm, appropriate additional sums of money, and embark upon a capital spending program to modernize the plant.[40]

These decisions prompted the formation of the San Juan Smelting and Mining Company, which acquired all the assets and liabilities of its predecessor. The directors included Henry Amy, Spencer Trask, George Foster Peabody, and others, nearly all of whom came from the two investment banking houses. They set the capitalization at $2,000,000 and used some of the stock to pay off the bonded debt and provide funds for new equipment. The directors also elected Henry Amy president, but he remained at his offices in New York while his son Ernest, a mining engineer with little or no experience in the smelting business, replaced Porter as manager of the works. But at the insistence of Palmer, Bell, and other large stockholders—all of whom had doubts about the ability of the Amys—the directors appointed Porter executive advisor. This was a strange position, and it was not to work out well.[41]

During the next two years, from 1888 to 1890, the Amys made improvements to the Durango smelter in hope of restoring its profitability. They installed new roasters, enlarged and modernized the blast furnaces, and purchased other machinery to improve efficiency in handling ores, bullion, and matte. Yet the plant continued to operate at a loss. The Amys could not overcome lower prices for silver and lead, high transportation costs, and energetic competition from the smelters at Denver and Pueblo.[42]

But as the nineties began, the San Juan enterprise seemed to become more prosperous. The capital spending program had lowered reduction costs significantly, and a new railroad line from Durango to the firm's coal mines had decreased the cost of fuel. The Denver and Pueblo smelters also chose this time to raise their treatment charges, which, said Porter, made them no longer competitive in the San Juan ore markets. Despite the improvement, a long-simmering feud between Porter and the Amys flared into the open in November 1890 when Henry Amy visited the works and slighted Porter. Porter decided to resign as executive advisor, but Palmer, Bell, and others persuaded him to remain because they still distrusted the Amys.

When Porter visited the smelter in January 1891, he described the outlook as splendid and predicted a profit of $60,000 for the year. But he also recommended more improvements, a suggestion that was well received in New York. The directors appropriated additional funds, but late winter snows in the San Juan delayed new construction for a time. The smelter also began to benefit from more energetic development work on mines at Telluride, notably those owned by the Smuggler-Union Mining Company, which had Porter as president and Hill, Pearce, Eddy, Grant, and Eilers on its board of directors.[43]

Despite Porter's predictions, during the summer the enterprise uncovered some financial chicanery in the upper levels of management. It was revealed in August that Edward Lambert, Jr., the firm's bookkeeper and the mayor of Durango, had embezzled more than $50,000 by drawing checks for freight charges above the actual amounts. The company recovered most of the money, but the affair prompted Palmer to suggest that the corporation should engage an outside auditor to examine the books at regular intervals, since he doubted that they had discovered all the underhanded activity going on in Durango and elsewhere.[44]

This incident blew over, and during the winter of 1891 the enterprise turned its attention to increasing the capacity of the smelter. Once the directors and eastern management had raised the requisite capital, Ernest Amy installed more roasting units to handle sulfide ores and two more blast furnaces to increase bullion production. The additions doubled the capacity, from one hundred to two hundred tons of mineral daily—courageous work for an enterprise that had rarely earned a profit. Yet Porter was still dissatisfied. He doubted that the roasting units would be adequate for working the huge quantities of sulfides coming from the San Juan mines, and he hoped that expected shipments from Rico would not require roasting. When Henry Amy arrived from New York in March, Porter convinced him to finance the construction of more roasting units.[45]

Porter also remained critical of the firm's cash position and the way Ernest Amy handled the ore supplies—interrelated problems. The enterprise never had enough working capital to carry what Porter regarded as an adequate quantity of mineral. The smelter acquired its iron flux from mines on Red Mountain, deep in the San Juan, but heavy winter snows put an end to railroad service until May. Nearly every year the Durango works came perilously close to exhausting its supplies, which would have left Amy with two undesirable alternatives: either curtail operations or else buy flux in Leadville at much greater cost owing to the freight tariffs generated by the roundabout transportation. Yet even more serious was the problem besetting everyone in the industry—securing adequate supplies of silver-lead ores high in their content of base metal. Amy never had enough working capital to negotiate contracts that might have provided the works with the mineral it needed. The smelter always remained under pressure operating from month to month and from shipment to shipment. Even so, Porter believed the enterprise had turned the corner into a long period of profitability. Yet his optimism was not shared throughout the firm. In January 1893 both William A. Bell and George Foster Peabody grew uneasy about silver prices, whose steady erosion had grown more rapid—a bad omen not only for the San Juan Company but for everyone in the industry.[46]

Since the middle of the 1880s, American smelters had imported ever-larger quantities of lead-bearing ores from Mexico. To do this not only was good business—mines south of "thirty-two" offered new markets—but also essential, for the minerals provided the lead needed to reduce dry ores mined in the West. Most shipments came into the United States through El Paso, Texas, the junction of the Mexican Central Railroad with the Santa Fe, Southern Pacific, and other American lines. Ores

The Durango smelter about 1900. Harry H. Buckwalter, photographer. Colorado Historical Society, Denver.

moving north through El Paso rose from a miniscule 1,800 tons in 1884 to a substantial 68,000 tons only four years later in 1888.

During this time the smelting industry functioned under the Windom customs decision made in the early 1870s. The tariff law required the federal government to collect a duty on all lead ores imported into the United States; but, under the Windom rule, if the value of gold and silver exceeded the value of lead, the shipment entered free as gold or silver ore, on which there was no duty. This decision enabled American smelters to avoid paying a tariff on minerals they brought across the border from Mexico. At first the Windom rule produced little rancor between miners and smelters, but controversy flared when Mexican ores began coming across the border in ever-larger quantities.[47]

As the price of lead declined and American smelters refused to increase their return, mineowners seized upon the idea of protection as a means of boosting their profits. They began their efforts to change the tariff in 1888. Their goal was simple and straightforward. They wanted Congress to eliminate the Windom rule and place a duty on the lead content of all ores brought into the United States. And through their petitions and lobbies they made their case. What had depressed the price of lead? The vast quantity of imported mineral excavated by cheap Mexican labor. And what had been the overall effect of imports? They had compelled American mines to curtail production and forego profits, they had deprived railroads of freight revenue, and they had thrown American labor out of work. This was a typical protectionist argument—neat, plausible, emotional—and the mineowners received a favorable hearing because their ranks were large.[48]

Most smeltermen opposed any change in the law and sought to make their views known to Congress. August R. Meyer, then emerging as one of the leading figures in the business, represented seventeen of the twenty-four large firms in the industry at hearings held by the Ways and Means Committee of the House of Representatives. Meyer gave his testimony on March 12, 1890. He pointed out that most firms like the Philadelphia,

St. Louis, and his own Consolidated Kansas City enterprise needed Mexican mineral in order to process dry ores mined in the United States. He urged the committee "not to accede to the demands of a minority of smelters and miners."[49]

The committee questioned Meyer closely. One member pointed out that both Dennis Sheedy and Arthur Chanute of the Globe enterprise as well as Guy Barton and James B. Grant of the Omaha and Grant firm had come out in favor of the proposed tariff. To this Meyer replied—weakly—that these companies would benefit immensely if his and other ventures went out of business. He added that he thought Grant and his colleagues were on the verge of changing their position. Meyer failed to mention, however, that both corporations owned lead-producing mines. Another member of the committee produced a letter from Anton Eilers supporting the proposed legislation, but Meyer pointed out that the letter was a year old and that Eilers had reversed his position. To prove it, he presented a recent letter from the smelterman to Congressman William McKinley, the committee chairman.[50]

Despite Meyer's efforts, the new McKinley Tariff, which became law on October 1, 1890, set a duty of one and one-half cents a pound, or thirty dollars a ton, on lead ores and a similar rate on the lead content of silver- and gold-bearing mineral. Since the market price of lead hovered around four and one-half cents a pound, the duty amounted to roughly one-third of the value of any shipment coming across the border. This eliminated whatever profit on lead American smelters could make by reducing Mexican ores that were also essential in processing the American product. Ironically, shortly after the new tariff became law, the market price of lead declined again and did not rise to the level of 1890 until the end of the decade.

The McKinley law made one saving concession to the smelting industry. It permitted individual companies to import Mexican ores in bond, reduce them to metal, then reexport the lead for sale outside the United States. This arrangement enabled American smelting companies to get the mineral they needed, but the extra transportation costs, lower international prices for lead, and the waste of capital tied up in bond combined to reduce or eliminate what profits might be made from the reduction of Mexican ores.[51]

The McKinley law had one unforeseen consequence. Although the new tariff walls ended the free movement of silver-lead ores northward into the United States, the statute accelerated the flow of American capital southward into all phases of the Mexican minerals industry. Such investment had been going on for some time, as the ores of Chihuahua, Durango, Zacatecas, and other states presented Americans with opportunities that were enhanced by the encouragement of the Mexican government. What the McKinley tariff did was speed the flow of capital coming through entrepreneurs in the smelting industry. It was no coincidence that beginning about 1890 they grew particularly active in developing Mexican mines and building silver-lead smelters that employed the techniques evolved in the United States over the past two decades.

One of the key people in this was Robert S. Towne, the short, red-bearded man who was second vice-president of the Consolidated Kansas City Smelting and Refining Company. Towne had been active in mining districts south of the border for several years, but in 1890 the president of Mexico, Porfirio Díaz, granted him what were known as concessions—five of them—to build reduction plants in the heart of the mineral country. Towne built no more than one, but this was a major smelter forming the keystone of a vertically integrated empire in metal.

Towne put his ideas into being in 1890 when he organized the Compania Metallurgica Mexicana, a New Jersey corporation despite its Spanish name. Towne held the controlling interest, but because of an arrangement much like Nathaniel P. Hill's with the Boston and Colorado Company, a minority stockholder—in this case Nathaniel Witherell—served as president and kept offices in New York. Towne then erected a huge smelter at San Luis Potosí, but this was not all. With the plant under construction during the winter of 1891, he acquired several important mining properties for the CMM, as it was known, and built the Mexican Northern Railroad to bring ores to the works. Much of the capital came through the investment banking firms of Spencer Trask in New York and Lee, Higginson in Boston, both of which had come to Towne's attention—and vice versa—through the offices of August R. Meyer, with whom Towne always worked closely. With so much money coming in through these houses, it was hardly coincidental that A. Foster Higgins of Spencer Trask held the post of treasurer in the "metallurgical company."[52]

Two other firms of lesser importance appeared during the nineties. One was the International Metals Company, organized in 1894 as a subsidiary of Meyer's Consolidated Kansas City enterprise. International acquired several mines, built a number of crushing and sampling works, and erected a small smelter at El Carmen. The other was the Velardena Mining and Smelting Company, organized as a subsidiary of the Omaha and Grant firm and soon to build reduction works at the town of Velardena. Neither venture had a great influence on the evolution of the smelting industry in the Rocky Mountains. Yet they gave the parent enterprises another dimension and fostered the export of capital and technology from the United States into Mexico.[53]

Yet, important as were these companies and their entrepreneurs, by far the most significant were the new men in the business—Meyer Guggenheim, his seven sons, and their investment vehicle, the firm of M. Guggenheim's Sons. They also perceived new opportunities in Mexico. On October 9, 1890, scarcely a week after the McKinley Tariff became law, Daniel Guggenheim, the heir apparent as head of the family business, obtained from Porfirio Diaz a concession to build three smelters to foster the development of the Mexican mining industry. The Guggenheims later created the Gran Fundicion Nacional Mexicana (the Great National Mexican Smelting Company), which erected works at Monterrey during the winter of 1891. A few years later the family organized the Guggenheim Smelting Company, which built a plant at Aguascalientes in 1895 and a refinery at Perth Amboy, New Jersey. By the end of the decade these works, neatly tied together through the firm of M. Guggenheim's Sons, made the family empire one of the foremost in the business, rivaling the Consolidated Kansas City, the Omaha and Grant, and the United firms as one of the dominant companies in the industry.[54]

In the meantime, as the major entrepreneurs battled over the tariff question and formulated plans to build smelters in Mexico, they also began their first efforts to unite the independent, competing firms into a single holding company. On November 30, 1889, Anton Eilers, Guy Barton, Alfred Geist (now with the Philadelphia firm), plus representatives of the Consolidated Kansas City venture and two other men got together at the Millard Hotel in Omaha. Their purpose, it was said, was to devise a method for protecting western miners and smelters from the "outrages" of the so-called Lead Trust, the newly formed National Lead Company, whose output had contributed to the low price of lead. These men discussed possible ways of organizing, but they failed to agree upon a definite course of action. Nevertheless, someone suggested that a single enterprise controlling the major western smelters could earn a profit of $4,000,000 annually, a return of about 40 percent on invested capital. The conferees agreed to meet again in February 1890. Meanwhile, Eilers was to visit the corporate offices of each enterprise, examine the plants and books, and make an assessment of every firm's property.

The next meeting took place on February 13, 1890, at the Palmer House hotel in Chicago. On this occasion nearly every firm in the industry sent a representative. The delegates discussed common problems and interests, and then at least a majority agreed in principle to organize the United States Smelting and Refining Company, a firm that would absorb all the plants and properties of the major processors: the Philadelphia, Colorado, Pueblo, Globe, Consolidated Kansas City, Omaha and Grant, and San Juan enterprises operating in Colorado and elsewhere; the Hanauer, Mingo, and Germania firms working in Utah; the Montana and the Helena and Livingston companies in Montana; the Chicago and Aurora venture of Illinois; and the old Balbach enterprise of New Jersey. Left outside the corporation would be the Boston and

Colorado Smelting Company, which never reduced silver-lead ores, and the St. Louis Smelting and Refining Company, which was now owned by the National Lead Company.

Despite such sweeping plans, the United States enterprise foundered on the rock of financial reality. The independent smelter owners wanted huge amounts of money for their plants and property, but the chief figures supporting the merger did not have enough capital to buy so many operations. The "trust" movement of the nineties had only begun, and the great financiers of New York as yet had little interest in uniting the individual smelting companies into one body corporate. And so the ambitious plans of the United States enterprise fell by the wayside in the course of the year.[55]

By 1893 it was no longer correct to speak of a Colorado industry, assuming that it ever had been—and of course it had not. But lured by profitability and pushed by necessity, the firms that centered their operations on Denver, Pueblo, Leadville, and Durango had transformed themselves from regional into national, if not international, processors. Some remained independent, others formed part of larger holding companies. Few small firms now remained in an industry dominated by large, integrated corporations. For them the future appeared bright, but lower, unstable prices for metal loomed unknown on the horizon. Over the next few years these and other developments would bring significant change to the structure of the business.

Chapter 8

Hard Times

THE YEAR 1893 OPENED ON AN OMINOUS NOTE. THE PRICE OF SILVER, in a slow downtrend since the close of the Civil War, declined further as winter evolved into spring. Then on June 26 came a thunderbolt. The Herschell Committee of the British House of Commons published its report recommending that Her Majesty's mints in India end the coinage of silver rupees. The news precipitated a frenzy of selling on the metal exchanges of the world. Within hours the price of silver plunged 20 percent, from eighty to sixty-four cents an ounce. Ores in the ground, mineral aboard railroad cars, bullion in smelting furnaces, and metal at refineries was suddenly worth only four-fifths of what it had been. There was chaos in mining and smelting circles. "Everything turned upside down at the low price of silver," declared one mineowner the day after the debacle.[1]

When the Herschell Report became public, Edward Eddy of the Omaha and Grant Smelting Company happened to be in London. Hearing the news, he cabled his associates that the price of silver had collapsed, and he predicted it would go even lower. When Eddy's message was received an ocean and a continent away, James B. Grant, Guy Barton, and Edward W. Nash, who were all in Denver, immediately held an impromptu meeting. Grant later remembered that everyone thought the firm could not do business in such an economic climate. And with that they decided to close the Denver and Omaha smelters until the price of silver stabilized at whatever level it might attain.

With market conditions in upheaval, other smelting enterprises followed the same course. Three days after Grant and his colleagues held their hurried meeting, W. W. Allen, general manager of the Pueblo Smelting and Refining Company, telegraphed Franklin Ballou's agency in Leadville: "Receive no ore for account. . . . Will neither unload it nor pay freight charges thereon until market conditions are fully settled." Two days later, on July 1, Sidney Bretherton, who managed the American smelter in Leadville, notified shippers that his firm would no longer make any ore purchases, but Bretherton continued reducing ores until he exhausted his supplies.[2]

A few enterprises tried to maintain operations in the face of the crisis. The Arkansas Valley plant cut the wages of its workforce in an effort to lower reduction costs, but this expedient did nothing more than precipitate a strike. Finally, on August 18, August R. Meyer telegraphed his local manager, Robert R. Rhodes to cease processing because of the "mob," the scarcity of ores, and the unfavorable market conditions. Only one smelter continued operations throughout this time—Nathaniel P. Hill's Argo works. The market value of the output fell off owing to the collapse of silver prices, but Hill and Richard Pearce kept the reverberatories working through the summer and fall because they had large quantities of ores bearing gold, whose worth held steady at twenty dollars an ounce.[3]

Smelting enterprises were by no means the only firms affected by the Herschell Report. Mining companies suffered in much the same way, and they also closed, throwing thousands of men out of work. The doors slammed shut in Leadville, Aspen, the San Juan—all across the high country—and in every western silver region—the Salt Lake Valley, the Coeur d'Alene, the Great Basin. The collapse and instability in the value of silver, plus the ravages of the Panic of 1893, which disrupted the national economy, brought the entire silver-lead industry to a virtual standstill. No one knew it then, but hard times would prevail for the rest of the decade, fostering turbulent labor disputes and promoting structural changes in the ore reduction business.

The silver-lead smelters remained closed for the rest of that troubled year of 1893, but this did not end the work of owners, managers, and metallurgists. No one expected the furnaces to remain idle forever. Despite the enforced hiatus, entrepreneurial activities continued. Smelters and miners haggled over assay values, prices for metal, reduction charges, and sites for umpires. Metallurgists redesigned furnaces, overhauled machinery, and planned improvements that would increase efficiency and maintain profit margins when the day came to set the smelting units back in blast. Purchasing agents kept in touch with mining companies, readjusted contracts, and discussed future ore shipments for whenever the time came to reopen the mines. Yet labor had a hard time. Except for handfuls of employees retained to man skeleton crews at the smelters, working people had to shift for themselves in a hard economic environment.[4]

With the smelters closed and silver depressed, some mineowners looked with new favor on the possibility of working their own ores, and perhaps the experience of the Aspen Mining and Smelting Company was typical. This enterprise had erected a plant in 1883 and reduced its own mineral for three years until the railroads made it more profitable to ship the ores to Leadville, Denver, and Pueblo. The firm took bids from all smelters and leased the Aspen works to the Hardinge Smelting Company, which operated sporadically and without much success until 1892. But, shortly after the collapse in silver prices, Jerome B. Wheeler, a founder of Macy's in New York and now president of the Aspen venture, directed his local manager Fred G. Bulkley to look into the possibility of reopening the smelter.[5]

Bulkley thought this an excellent idea—or so he wrote Wheeler. Bulkley found he could get iron flux in Leadville and the copper he needed for producing matte in Salt Lake City. He also learned that freight rates on bullion had fallen considerably since the eighties when the company had shipped its furnace product to Omaha for refining. But despite such propitious findings optimism proved short-lived. After a few weeks of investigation, Frank Bulkley wrote that he and his brother were "not sanguine" about the prospect of operating the smelter profitably. When Wheeler arrived in Aspen later in July, all three concluded that it would be prudent to leave the furnaces out of blast and continue selling ores to the smelters whenever they reopened for business.[6]

And during the fall Grant, Meyer, and others readied their plants to resume operations. Silver steadied itself at something above sixty cents an ounce, and after a time both miners and smelters grew confident the price would not plunge further. With that, mining companies engaged new work forces—at lower wages—and smelting firms did likewise. By January 1894, with mineral again moving toward their plants, the smelters blew in their furnaces and resumed the production of bullion and metal. But now there were changes that soon grew apparent to smelterworkers, mineowners, and ore buyers.

A few weeks before they resumed operations, the major smelting firms had sent representatives to Denver to discuss issues of common interest. They talked about establishing uniform reduction rates throughout the industry to eliminate what many referred to as "ruinous" competition. They also wanted to pressure the railroads into granting lower freight rates, something they might well do by acting collectively. Negotiations went on through the winter months, and by February the major processors had agreed to advance treatment charges an average of two dollars a ton and purchase certain classes of ore through a small group of favored brokers. To run the system, they established the Smelters Association, which was to operate through a clearinghouse in Denver.[7]

The results became apparent when the arrangement went into effect in February 1894. In June, Frank Bulkley of the Aspen Company wrote that competition for ores had vanished. Later in the year he calculated that his reduction costs had risen about $2.80 per ton, partly because he found his new production more refractory than hitherto, but largely because the Smelters Association had made a uniform increase in treatment charges.

Unfavored ore buyers found themselves in dire straits. Scarcely two months after the clearinghouse came into being, John B. Henslee of Denver wrote that the combined effects of the pool and low metal prices had caused the "annihilation" of 90 percent of his business. Franklin Ballou's agency in Leadville suffered a similar fate. Because the Bi-Metallic Smelting Company, a new enterprise in Leadville, had decided to remain outside the association, the smelters switched their purchases of iron flux to other firms. Ballou's sales fell off sharply.

Despite its apparent sucess, the Smelters Association did not have a long life. In February 1895 the Guggenheims decided to withdraw. After a single year of operation, the clearinghouse came to an end, like most informal pools in American industry. The smelters resumed their sharp competition for ores.[8]

Even though the association failed, the smelters still managed to establish a new purchasing policy throughout the industry. For years they had "settled" with miners and ore brokers on the basis of the value of metal prevailing on the day of shipment, but after the Herschell Committee made its shattering report, the price of silver fluctuated violently. It could rise or fall several cents an ounce between the day of shipment and the day the smelters could sell the bullion or refined metal. To increase stability, the smeltermen adopted the practice of paying mineowners on the basis of thirty-day or forty-five-day futures, the time required to ship, reduce, refine, and market silver. Thus the price of metal prevailing thirty or forty-five days after the shipping date would determine the settlement, a practice that eliminated windfall profits and spot losses that might accrue if the value of silver shifted in the interim.

Finally, plant managers had to lower their smelting costs, and so, as they hired new work forces, they compelled their employees to accept reductions in pay ranging from 10 to 15 percent of the former rates. The new scales did not equal the 25 percent cut that miners had to accept, but smelter workers still found their standard of living on the decline. This reduction in pay became a major cause of the labor unrest that wracked the industry in future years as workmen sought to regain the wages lost in the winter of 1893 and 1894.[9]

These changes notwithstanding, some plants never reopened for business, the chief losses coming in Leadville, as might have been expected. The Harrison Works had ceased operation just before the Herschell Committee issued its report. The steady erosion in silver prices, along with technical obsolescence, had made the plant unprofitable for some time. The traumatic events of 1893 did nothing except worsen the situation, and with that the National Lead Company, which had bought the St. Louis firm, decided to close the works forever. Leadville's first smelter had come to the end of the line after fifteen years of operation.

Yet this was not all. Another casualty was the second largest plant in camp, the American works of the Chicago and Aurora Smelting and Refining Company. After silver collapsed in June 1893, Sidney Bretherton, who had managed the smelter for a decade, continued running but only to process the ores he had in stock. In September he suspended operations, throwing four hundred men out of work. An old plant and obsolescent technology now forced the hand of the parent firm. Rather than appropriate the capital needed to renovate the smelter, Henry I. Higgins and his associates decided to withdraw from the Leadville industry in favor of their plants in Illinois. The small Elgin works reopened briefly in 1894, but only to treat the ores on hand before closing once more. This left only the Arkansas Valley and the new Bi-Metallic and Holden smelters remaining in Leadville, which only a decade before had dominated the western smelting industry.[10]

Despite the demise of the American and Harrison plants, the Consolidated Kansas City Smelting and Refining Company was one firm that intended to remain in business in the central Rockies. August R. Meyer and his associates used the months of enforced idleness to make changes in management and renovate the Arkansas Valley smelter. Meyer brought in Joseph H. Weddle to replace Robert R. Rhodes as manager, and Weddle overhauled the blast furnaces and made other improvements. Meyer and his ore buyers negotiated new contracts with mining companies, and as fall drifted into winter the ore bins at the plant came to bulge with ten to twelve thousand tons of mineral.

As the smelter made preparations to resume work, Weddle engaged a new force of workmen, but at wages 10 to 15 percent below what they had been.Then on January 16, 1894, he put the works into operation on a limited scale. Once under way, he increased the number of employees to about 250, about two-thirds the normal complement, and by February the furnaces were smelting about 350 tons of ore a day. One observer noticed an "air of old-time activity" down at the works.[11]

Yet Meyer and Weddle found their operations handicapped by the low prices of silver and lead. Even though they had cut the wages of their employees, lowered reduction costs, and increased smelting charges, their profit margins still remained narrow. Weddle told one reporter that the price of silver had reached such a low point that miners found it difficult, it not impossible, to produce certain grades of mineral that had once netted them a profit. Because many shippers could not work under the prevailing values, the Arkansas Valley works had difficulty in obtaining essential classes of ore.[12]

But Meyer and Weddle were resourceful, and during the nineties they reached near, far, and wide for the mineral they needed to keep the AV furnaces in blast. They bought huge consignments of ore from the Yak, Ibex, Lee Basin, Denver City, and other mines in Leadville, and they competed for the production of Aspen, Breckenridge, and other towns. When the virulent miners strike of 1896 convulsed Leadville, ending production and forcing other smelters to close, Meyer and Weddle had their ore buyers scour other towns for flux, carbonates, sulfides, and silicious minerals. The Arkansas Valley smelter remained in operation. No western mining district was too remote for Meyer's agents. In 1897 one man ventured all the way to British Columbia in search of contracts. Yet lead-bearing ores continued to be in short supply, and like other smeltermen Meyer and Weddle brought in what they needed from Mexico and from the Coeur d'Alene in Idaho. On those occasions when Weddle fell seriously short of base metal, Meyer had pigs of lead shipped from the refinery at Argentine, Kansas, back into the central Rockies.[13]

Despite their unfavorable location on the North Arkansas,

Meyer and his colleagues did what they could to improve efficiency and lower their unit reduction costs. They used the summer of 1894 to install a horseshoe furnace to augment their roasting capacity. Three years later, keeping in the forefront of technological advance, they had Weddle erect a building to house what was known as a Ropp straight-line roasting furnace. And, to keep pace with the increasing production of sulfide ores, Meyer and his associates bought one of the defunct smelters in camp. These improvements and the economies generated by smelting ores on a large scale permitted Meyer and Weddle to lower their reduction costs from $5.87 per ton in 1892 to $3.98 in 1897, a decrease of 32 percent. This was a fine achievement, but the Arkansas Valley works still had the highest unit costs of the plants owned by the Consolidated Kansas City firm.[14]

Meyer and his colleagues financed their improvements through the retention of earnings and the sale of securities. In January 1894 they issued $1,000,000 in preferred stock, paying cumulative annual dividends of 7 percent. Two years later they sold another issue, but it took longer to market the shares. Although the firm acquired large amounts of money this way, its demand for cash remained voracious. In 1897 Meyer wrote Henry Lee Higginson, a Boston financier, that he was "tied hand and foot without it—can pursue no policy long enough to get results and am at the mercy of accident." An overstatement, perhaps, as Meyer was given to this, but the words reflected his need for money in a capital-intensive industry.[15]

Despite the perennial quest for cash, Meyer remained proud of his achievements as president of the firm. His work in raising capital, plus the economies generated at Leadville, El Paso, and Argentine, sustained the enterprise during the hard years of a hard decade and augured well for the future. And Meyer knew it. In 1897 he wrote Higginson that "we have . . . an absolute patent upon the success of our business because we have relied entirely and only upon rational and logical causes for success, namely, cheap operating." Meyer and his associates used the money saved to pay principal and interest on the corporate debt, as well as dividends on the preferred stock. Yet the firm could not do so well by its common stockholders. After the price of silver collapsed, they received dividends—small ones—only in 1895, 1896, and 1898.

Miniscule returns to the common shareholders notwithstanding, Meyer looked to the future with optimism. His firm held an excellent position, with three smelters, a refinery, sampling agencies, mines, and Mexican subsidiaries. And he expected ever-increasing ore shipments coming north from Mexico, if for no other reason than that A. Foster Higgins, a director of the enterprise, was building a new railroad south from El Paso along the Sierra Madre. In May 1897 Meyer wrote that "the outlook for the business is better now than it ever was and . . . the good outlook is permanent." His company was profitable despite "low prices" for metal and "excessive competition" for mineral. He was also confident that upcoming talks between the major smeltermen would lead to higher reduction charges throughout the industry and greater profits for the Consolidated Kansas City enterprise.[16]

Despite the long, slow decline of the reduction business on the North Arkansas, some entrepreneurs still saw Leadville as a good site for smelting ores. This may seem paradoxical, for the obvious trend had been to build new plants on the Great Plains; but the mines and dumps seen everywhere in camp held vast quantities of low-grade minerals that could not be shipped anywhere for profitable reduction. The people now wishing to enter the industry thought that a relatively new technique known as pyritic smelting might make it profitable to reduce these materials. This process employed huge blast furnaces to reduce low-grade copper and iron sulfides into a matte that would collect and hold the precious metals. The key was the low cost of fuel. Ideally, a very small amount of coke, perhaps 1 to 5 percent of the ore charge, would ignite the sulfides, which would then generate the heat required for their own reduction.[17]

While many metallurgists held doubts about the feasibility of the process, no one could deny that it had potential, and this attracted a group of prominent Colorado mineowners. The chief figures were Eben Smith, a tall, thin mining man whose career stretched back to the boom days at Central City, and his associate Franklin Ballou, a Leadville ore-buyer who sold large quantities of flux to the valley smelters. Curious about the method, Smith and Ballou discussed the matter with T. S. Austin, a coinventor of pyritic smelting, made some rough calculations, and thought the project over. Eventually they concluded that they could build a plant in Leadville, process the local ores (some of which would be their own), and sell the matte to the Argo works, the Pueblo Smelting and Refining Company, or refinieries in the East. From this they would make a profit of $2.50 per ton.

The scheme seemed lucrative, but to raise capital and spread the risk Smith and Ballou convinced three entrepreneurs to join them in the venture. These men were David H. Moffat, Jr., a banker, railroad builder, and mining speculator known throughout Colorado; G. E. Ross-Lewin, another banker and mineowner; and James C. Wigginton, still another mining man. Once they had agreed in principle and detail, these five organized the Bi-Metallic Smelting Company, electing Smith president and Ballou general manager.[18]

Rather than erect a new plant, the Bi-Metallic enterprise purchased the works of the defunct La Plata Mining and Smelting Company. Clearly, Smith and his associates intended to minimize whatever losses they would have to absorb if the pyritic process failed to meet their expectations. Ballou renovated the works, installed new machinery, and on July 30, 1892, set in blast a huge furnace with an ore capacity of one hundred tons daily. This was an experimental unit, but Smith and his colleagues were so encouraged by its success that in the fall they installed two more furnaces. Simultaneously, Ballou bought large quantities of mineral from mining companies in Leadville, Breckenridge, and Aspen, and some from the new camp of Creede, on the eastern flank of the San Juan.[19]

As the year 1893 opened, Smith and his colleagues held high hopes for their new venture; but like other reduction men they had to close the plant when the price of silver collapsed and wait for the value to stabilize before resuming operations. This came in January 1894, as it did for other firms that decided to continue working. Ballou engaged a new smelter crew—at lower wages—then set one furnace in blast at the end of the month.

Over the next few weeks, however, he found himself plagued by shortages of certain ores essential in preparing the optimum reduction charges. Nonetheless, he blew in all his furnaces, and by May he had the plant smelting three hundred tons of mineral daily.

Yet a lack of silicious ores continued to hamper Ballou's operations for the rest of the year. When he set the furnaces back in blast, he had anticipated daily shipments of at least seventy-five tons of this mineral from the Amethyst mine across the mountains at Creede. But the ores were not forthcoming. When the mine closed in 1893, water had flooded the shafts and stopes. Obviously it had to be pumped out before full-scale mining could resume, but when Smith and Ballou set to work, the pumping went more slowly than expected. This miscalculation retarded production for the smelter, and Ballou had to shut down one furnace that summer until he could find an alternative source of silicious ores in Breckenridge and other camps.

Yet this was only one problem facing the enterprise. Because Smith and Ballou failed to take the Bi-Metallic Company into the Smelters Association, they found themselves excluded from the lower freight tariffs their rivals negotiated with the Rio Grande and other roads. The plant also had higher operating costs than its competitors because Smith and Ballou were still using the narrow-gauge tracks built for the old La Plata Company more than a decade before. Smith, Ballou, and a group of Leadville mineowners tried to persuade the railroads to lower their rates and widen the roadbed, but the talks brought no immediate relief.[20]

Despite these problems the smelter remained in blast for much of the decade, although it never returned the profits envisioned in 1891 when silver brought higher prices. Ballou drew upon the production of Leadville, Creede, Aspen, Breckenridge, and other camps, reduced ore to matte, and sent the output east for refining. On one occasion he found himself in the unusual situation of having purchased a consignment so rich in gold that the pyritic process, designed for low grades of mineral, failed to recover the metal. Ballou had to ask the shipper to send this product elsewhere and provide him with poorer material.[21]

These efforts notwithstanding, the Bi-Metallic smelter could not remain in operation in the face of the great miners' strike that convulsed Leadville in 1896. Ballou kept his furnaces in blast as long as he could by using his stockpiles and drawing on the production of other camps, but he finally had to close when his own workmen walked off the job in sympathy with the miners on the picket lines. The shutdown was short-lived, as the smelter workers soon came back, but the plant depended upon Leadville for the bulk of its mineral. When Ballou exhausted his main supplies, he had to close the works again. It remained this way until the National Guard, called out by the governor, ostensibly to keep order, broke the strike during the hard winter months. In the interim Ballou sold what matte he had to the Arkansas Valley smelter and tried to sell his excessive stocks of iron flux to the Colorado Smelting Company. When the strike came to its bitter end the mines resumed production, and in July 1897 the Bi-Metallic plant set its furnaces back in blast.[22]

The collapse in silver prices hindered the operations of Smith and Ballou, but for other entrepreneurs using the pyritic process in Leadville, the fall in values was a disaster. Back in the winter of 1891, as Smith and company formulated their plans to enter the industry, another group of investors were thinking along the same lines. The leading figure was that master promoter of smelting ventures, the ubiquitous Edward R. Holden, now anxious to enter the business a third time. He never seemed to have much trouble raising capital for such enterprises, and this one proved no exception. He obtained financing in Denver, then he and his new associates organized the Holden Smelting and Refining Company. They broke ground in Leadville on April 16, 1892, pushed construction in the spring, and set two one-hundred-ton smelting units in blast in June. Holden encountered a number of mechanical problems with his giant furnaces, and operations went on sporadically for several months until his metallurgists ironed out the technical deficiencies.[23]

Early in 1893, Holden—ever the dreamer, glad-hander, and promoter—announced plans to erect two more smelting units that would make the plant the largest pyritic smelter in Colorado. He even signed contracts to reprocess the slag lying on the dumps of the old Grant and La Plata smelters, but the collapse in silver prices struck the firm like a thunderbolt. The smelter had never made enough money for Holden to meet the payment on loans he had secured in launching the venture. He had kept his creditors at bay, but once the price of silver plummeted in June, they came after him like hungry wolves. In July they attached the works for $175,000. Holden tried to keep his furnaces in blast to generate cash, but when the railroads refused to deliver more coke on credit he had no choice but to suspend operations.

Lacking any alternative, Holden had to sit down to discuss finances with the men who had lent him money. After a round of talks it was agreed that Holden would reopen the plant in conjunction with a committee representing the creditors. Everyone hoped to resume operations that fall, but when the mines remained closed Holden could not put his furnaces back in blast. At this point the bankers foreclosed, and in January 1894 they sold the smelter at a sheriff's sale. Edward R. Holden's third and last effort in the smelting industry had come to an end.[24]

Yet if Holden was through, the plant was not. After the sheriff's sale it passed to a group of Denverites who organized themselves into the Union Smelting Company. These men included R. W. Woodbury, who had once tried to work through Horace Tabor in persuading the Arkansas Valley Company to relocate in Denver, and William R. Harp, who had once started a short-lived smelting firm that erected works at Canon City. With their associates they set the capitalization at $250,000 and named Harp general manager. Woodbury and Harp intended to commence operations in February 1894, but repairs to the plant, ore shortages, and inadequate supplies of coke delayed the start until March. Then Harp blew in two furnaces and added a third unit later in the year. Yet he could never obtain the ores he needed to prepare optimum furnace charges for reduction. Profits proved elusive, and the plant closed in December.[25]

Despite the setback, Harp and his colleagues refused to withdraw from the industry. Later that month they reorganized the firm and obtained new capital from the Bank of Commerce in

Kansas City. Harp resumed operations, but soon the bankers decided to lease the smelter from the Union Company. They continued working until July 1896, when the miners' strike curtailed ore production and forced the plant to take its furnaces out of blast. When the lessees declined to renew their contract, the sheriff of Lake County again took possession of the smelter and sold it to satisfy a deed of trust.[26]

This time the purchaser was Absolam V. Hunter, a prominent Leadville banker associated with the mining ventures of August R. Meyer and Henry Lee Higginson. Hunter paid a reported $56,000 for the plant, then tried to sell it to other smelting enterprises, but without success. It lay idle for two years until 1898, when Hunter persuaded Meyer to purchase the works to augment the roasting capacity of the Arkansas Valley smelter, which was confronted by ever-larger tonnages of sulfide ores. Weddle put the roasting units into operation, but neither he nor Meyer had any plans to activate the smelting furnaces.[27]

While the smelters in Leadville struggled along during this turbulent decade, significant changes befell the San Juan Smelting and Refining Company, the largest firm in southwestern Colorado. John A. Porter, the executive advisor, had predicted that 1893 would be a profitable year, but the cataclysm that destroyed the price of silver not only shattered his prognostication, but also forced the plant to close. It sat idle for six months until January 1894, when Henry Amy, the local manager who still disdained Porter, put the furnaces back into blast. Yet this proved premature. Many mines in the San Juan country were still closed, partly because of low metal prices, partly because of winter. Amy could not get the ores he needed, and in April he had to lay off workers and reduce the scope of his activity.[28]

With the smelter running at a loss, Amy and corporate officials in New York had to raise new capital if they hoped to remain in the industry. After several hurried meetings, the directors agreed to market $500,000 worth of bonds through the firm of Henry Amy & Company and to secure the debt through a mortgage on the Durango works and other property. Despite the adverse economic climate in the financial world, Amy & Company managed to sell the bonds, apparently at a discount. Yet the enterprise remained in poor financial health.[29]

Then came new developments. Over the years the major stockholders had invested large sums of money with little or no return, since the plant had always been a marginal processor. Now in the winter of 1894 and 1895 many wanted to get out before they lost everything, as there was no reason to expect the enterprise could ever run at a profit with the prices of silver and lead so much lower than ever before. Amy and his associates had no trouble finding buyers. The works had the potential to dominate the San Juan trade, and right at hand was John A. Porter, president of the Smuggler-Union Mining Company, which included Hill, Pearce, Eddy, Grant, and Eilers among its directors. What was more, Henry Amy had worked with James B. Grant as early as 1878 when the banking house had sent him to investigate the newfound ore deposits on the North Arkansas.

During the winter Amy and his associates opened negotiations with the owners of the Omaha and Grant firm, and perhaps others as well. In March 1895 the Durango press reported rumors that some sort of deal lay in the offing. No one knew what. Then in April came hard news. The directors of the San Juan Company announced that they had leased the plant to the Omaha and Grant enterprise for an indefinite period. And although no one said so, the agreement apparently included a clause that permitted Barton, Grant, and their associates to buy the company. Less than a year later the Omaha and Grant enterprise took possession of the San Juan firm as a wholly owned subsidiary.[30]

Once in control, Grant and his associates replaced Ernest Amy as plant manager with Franklin Guiterman, a man well known in mining and smelting circles. A graduate of both the Columbia School of Mines and the famous Bergakademie, he had already served as an assayer for the old Little Chief firm in Leadville, as superintendent of the Mingo smelter in Utah, and most recently as the manager of the Denver clearinghouse of the Smelters Association. Guiterman assumed his new responsibilities at Durango in April 1895, scarcely a month after the Omaha and Grant Company took over the plant.[31]

Barton, Grant, and Guiterman moved to strengthen their competitive position in southwestern Colorado. According to published reports they intended to renovate and enlarge the Durango smelter in hope of controlling not only the production of the San Juan, but also that of surrounding regions. With an infusion of capital, Guiterman erected new sampling works, larger ore bins, and more machinery. He also competed more energetically for San Juan minerals that had been going to Pueblo because of the turmoil in Durango. But his efforts sustained an unfortunate setback on September 25 when a fire at the smelter destroyed the new ore bins and other structures. Only by chance did the entire plant escape destruction. Guiterman estimated the damage at $75,000, a portion of which was covered by insurance. This setback proved only temporary, however. Once operations resumed, Guiterman established the Omaha and Grant Company as a solid competitor in the San Juan trade and himself as one of the chief figures in the enterprise.[32]

The silver crisis enabled Grant and Barton to increase the scope of their operations, but it had an opposite effect on Nathaniel P. Hill and his associates in the Boston and Colorado Smelting Company. The production of the Argo works fell sharply, from $6,062,000 in 1892 to $2,360,000 in 1895; there it stagnated for the rest of the decade. The low price of silver accounted for much of the decline, but other changes also contributed. During the nineties new smelter construction in Utah eliminated one of Hill's ore markets, as local buyers outbid buyers from Denver. Grant, Meyer, and other smeltermen also chose these years to erect electrolytic copper refineries to process their own matte. These plants eliminated a source of business long profitable for the Argo works.

Hill and Pearce also found their shipments of ore and matte from Montana declining. Before the report of the Herschell Committee, the Argo works had received large consignments of copper products from Butte, where Henry Williams ran the closely allied Colorado Smelting and Mining Company. He provided Hill and Pearce with much of the copper they needed to operate the Argo smelter, which could never get enough base metal from the high country alone. But after 1893 the lower prices for silver and copper combined with "excessive" freight tariffs to eliminate the profit made in shipping the products from Butte to Denver.[33]

Late in the decade Hill and his colleagues decided to sell their interest in the Montana firm—no easy task, for their holdings were large and the shares not listed on any stock exchange. In the winter of 1896 Henry Wolcott went east to arrange a private sale in Boston, always an important market for copper securities. He talked with Henry Lee Higginson, who was closely connected to the famous Calumet and Hecla Mining Company of Michigan. Higginson had the old mining engineer James D. Hague inspect the properties in Butte. Wolcott was so confident the negotiations would succeed that he had Hill, Pearce, and others endorse their stock certificates and send them to the Old Colony Trust Company in Boston, but in the end the deal fell through.[34]

Hill and his colleagues now turned their attention to New York. Wolcott made his first effort to sell the stock there just about the time a small coterie of financiers headed by Henry H. Rogers and William Rockefeller, both of Standard Oil, along with Marcus Daly and F. Augustus Heinze of the Anaconda Copper Company, were formulating plans to take control of American copper production, which represented about 55 percent of the world output. The scheme was to unite all the producers in Butte—and probably elsewhere—into a single firm that would market copper through another enterprise controlled largely by Rogers and Rockefeller. This corporation, to be known as the United Metals Selling Company, would regulate the international price. How Wolcott made contact with Rogers and Rockefeller remains unknown, but contact he made. Negotiations ensued, and in 1898 the Argo smeltermen sold their stock in the Montana firm. Shortly after, the Colorado Smelting and Mining Company became one of the first enterprises acquired by the Amalgamated Copper Company, which eventually took control of the Montana industry.[35]

With the business of the Argo smelter contracting as the nineties struggled forward, Hill and his associates took the unusual step of reducing the firm's capitalization. It had stood at $1,500,000 since Pearce, Crawford Hill, and Samuel F. Rathvon had reincorporated the firm in Colorado on April 21, 1892. But in October 1895 the stockholders held a special meeting in Argo, where they voted to decrease the figure to $750,000. Hill later explained to a congressional committee that the value of the plant and the money employed as working capital exceeded $750,000, but that the lower capitalization more accurately reflected the diminished scope of the smelter's activity.[36]

As the nineteenth century drew to an end, Hill's lifetime was also coming to its close. By February 1899 his doctors permitted him to spend less than two hours a day at his offices because of the ravages of a stomach disorder, but this failed to help his condition. The disease progressed, and his deteriorating health finally confined him to bed. He died at his home in Denver on May 22, 1900, thirty-six years after his first journey to Colorado.[37]

Even as Hill saw his business decline, his rivals in Denver and Pueblo continued to process huge tonnages of ore. The towns of Leadville, Aspen, Cripple Creek, the mines across the San Juan mountains, and districts scattered here and there in the Rockies sent their minerals down from the high country or across the plains to the reduction furnaces built by Eilers and Geist and Grant and all the rest. Most shipments consisted of sulfides or concentrates, since carbonates had virtually disappeared, and no one located any extensive deposits of mineral high in lead that the smelters might mix with dry ores. The major processors remained dependent upon shipments from the mines of Mexico and the Coeur d'Alene.[38]

By the later 1890s four large smelting empires had come to dominate the American industry. The United Smelting and Refining Company, which included Anton Eilers in the managment, had the most northern orientation, with plants at Pueblo, Great Falls, East Helena, and Chicago. The Omaha and Grant Company had a more central thrust, with works at Denver, Durango, Omaha, and Velardena. Meyer's Consolidated Kansas City firm had a more southwestern interest, with plants at Leadville, Argentine, El Paso, and El Carmen. Meyer

Guggenheim's holding company had the broadest and most southern orientation, with plants at Pueblo, Perth Amboy, Monterrey, and Aguascalientes.

Each of these firms enjoyed the benefits of horizontal and vertical integration. All owned mining properties, sampling works, several smelters, a refinery, and marketing agencies. All were technologically advanced for during the decade each had added equipment to electrolytically refine the copper matte evolved in the reduction of what were primarily silver-lead ores. All had marketing agencies that tried to differentiate their product by selling "Omaha lead," "Kansas City lead," and "Perth Amboy lead." Meyer and his associates even created their own "alchemist brand" of litharge, vitriol, and zinc sulfate.[39]

These firms and smaller rivals like the Globe Smelting and Refining Company had enjoyed a large measure of success over the years, but throughout the nineties the industry found itself plagued by low, unstable prices for silver and lead and what the chief entrepreneurs regarded as excessive competition. By 1897 many smeltermen favored a change in the status quo. Ideas ranged from a restoration of the old Smelters Association to the apportionment of competitive territory to the formation of a holding company. August R. Meyer wrote in May that "all along the line" he saw "a great desire to reach an agreement and raise margins." In the fall he and Daniel Guggenheim went to London to persuade the House of Rothschild to buy and sell the entire silver product of American smelters, thus raising and steadying the price. Yet their plan fell through.[40]

The desire to create greater stability throughout the industry prompted a meeting held that November at the Brown Palace hotel in Denver. Nearly every firm sent a representative, not just those working the production of Colorado. By prearrangement the delegates had kept the topics of discussion secret, but the news gradually came to light. The smeltermen had several immediate objectives: they wanted to arrange a new division of ores from certain districts in Colorado, eliminate conflicts that had arisen between individual enterprises, and end competition in a few mining districts. Beyond this, they talked about doing away with the old system of using Western Union quotations to buy mineral and sell metals. They also discussed the possibility of setting limits on the grades of Cripple Creek ores that would go to cyanide mills, whose efficiency and low costs were drawing shipments away from the furnaces at Denver and Pueblo.

That side of the talks dealt with immediate solutions to short-term problems, but the delegates took up a question of far greater portent. Everyone at the Brown Palace knew that for nearly two decades—perhaps more—informal pools like the old Smelters Association had failed and that the agreements they were negotiating would probably come to the same end. But this time another topic surfaced. The smeltermen discussed the possibility of uniting their enterprises into a single holding company that could raise treatment charges, increase profit margins, and guarantee stability where a loose organization could not. The convention provided a forum for ideas and proposals, but it failed to produce any agreements.

The smeltermen met again that month in New York City, and this time they hammered out a workable arrangement. Since the Denver meeting, each delegate had discussed the various proposals with the officers and directors of his company. Now they agreed to a partial division of territory that would permit a more even distribution of certain classes of mineral, and they created a new pay schedule to serve as a guide in negotiating with ore producers. The delegates also discussed the possibility of organizing the independent firms into a holding company, but again they could not reach a conclusion because their views diverged and they realized that no one man or firm had enough capital to buy the ores, smelters, refineries, bullion, and metals of the others. As a result of this meeting, however, the smeltermen gave Edward W. Nash, secretary and treasurer of the Omaha and Grant Company, the responsibility for visiting the corporate offices of each firm, examining what data might be available, and determining the willingness of the major stockholders to sell their interests to a holding company.

Talk about a resurrection of the Smelters Association and rumors about the creation of a "trust" stirred the mining community to action of its own. As hearsay and speculation swirled through the high country, Henry I. Higgins, secretary of the Leadville Mine Manager's Association, called a meeting to be held at the Brown Palace hotel. Once in session he and his cohorts discussed the prospective action of the smelters, but Higgins's group could take no action. There were too many mining companies for the group to agree on a single policy regarding the smeltermen, and as yet the reduction firms had not announced any changes in ore processing.[41]

Despite the incipient opposition from the mining industry, the smeltermen put their new agreement into effect early in 1898; but, like the Smelters Association and innumerable other pools worked out by these men from time to time, it fell apart in a short while. By this juncture, the chief people in the business had perceived that only a single corporation controlling the independent plants could guarantee an effective division of key ore markets and eliminate the competition that made for narrow profit margins and sometimes outright losses. But to create a holding company the smeltermen needed huge quantities of capital to buy up the smaller enterprises. Such funds none of them had.

Yet the industry was now to be carried along in the wave of late-century mergers, specifically by the move to unify the copper business. Early in 1898 Rogers, Rockefeller, and Lewisohn had organized the United Metals Selling Company as part of their plan to control the international price of copper. By this time nearly all the large, integrated silver-lead smelters produced substantial quantities of matte as a by-product of their main operation. Rogers, Rockefeller, and Lewisohn knew that if they could not control the copper production of the silver-lead companies, the United Metals firm could never achieve its objective. Thus they agreed to help finance a consolidation of the independent smelting firms in return for a contract to sell the copper that the combine would produce. As skillful manipulators of securities, Rogers and his colleagues also recognized the potential for profit in organizing a silver-lead combination.

Assured of financial support, Nash pursued his work in earnest. In 1898 he traveled around the country to persuade the presidents, officers, and directors of many firms of the advantages of selling out to the proposed combine. Wherever he went, Nash received a cordial welcome, for his reputation was wide and many investors were interested in making a profitable exit from an industry that seemed to have a speculative future. By 1899 most smeltermen had agreed in principle to sell their plants and properties to a syndicate that would organize a holding company.[42]

Then came the work of Moore and Schley, a New York investment banking house with much expertise in forming such combinations. Once Rogers engaged them, the financiers sat down to discuss what type of package would induce the stockholders of the independent companies to sell out. The underwriters eventually decided upon what appeared to be a lucrative offer. They would create a company that would issue equal amounts of common and preferred stock to be exchanged for the securities of the individual smelting enterprises. These were the terms. First, the underwriting syndicate—meaning Moore and Schley—would rate the plants of each firm at something above their real value. Then, for each $1,000 worth of property, the underwriters would offer ten shares of preferred stock, each with a nominal value of $100. As a bonus for selling, they would add another seven shares of common stock. Moore and Schley would keep the remaining three shares for their own expenses, commission, and profit.[43]

With the stage set, the major entrepreneurs in the industry went to New York to bargain with E. R. Chapman, the broker Moore and Schley had engaged to handle the final negotiations. Grant, Meyer, Eilers, and others found the inducements of the syndicate hard to resist. Chapman gradually assembled enough options to control the outstanding stock of nearly all the principal firms, and this persuaded minority shareholders to sell. Through these transactions the underwriters acquired options on the Consolidated Kansas City, Omaha and Grant, and United companies—an important threesome because the new enterprise would draw its management largely from them. Then came other ventures: the Globe, Pueblo, and Bi-Metallic in Colorado; the Germania and Hanauer in Utah; the Pennsylvania Lead Company, with plants in Pittsburgh and Salt Lake

City; and the Chicago and Aurora enterprise with works in Illinois and Colorado, although the American smelter in Leadville had not reduced a ton of ore since 1893.

On April 4, 1899, Moore and Schley obtained a New Jersey charter for the American Smelting and Refining Company. Rogers and his associates set the nominal capitalization at $65,000,000, divided into $32,500,000 of common and an equal amount of preferred stock bearing an annual dividend of 7 percent. The enterprise called in the options and took possession of most of the plants in the industry. In doing this, however, the company issued only $27,500,000 worth of each class of stock. The first management placed the remaining $10,000,000 in securities in the corporate treasury.

The new enterprise—ASARCO, as it came to be known—drew its officers and directors primarily from the three major smelting companies as well as the underwriting syndicate. On the board sat Nash, Barton, and Grant of the Omaha and Grant Company; Meyer, Witherell, and Towne of the Consolidated Kansas City venture; Eilers, Barton Sewell, and Walter S. Gurnee of the United group; and Rogers, Lewisohn, and Grant B. Schley of the underwriting concern. The directorate also included Dennis Sheedy, Mahlon D. Thatcher, and David H. Moffat, Jr., but they resigned in a short time. Nash was elected president partly as a reward for his efforts in forging the combination. Sewell and Thomas M. Day, Jr., took office as vicepresidents; Winthrop E. Dwight was treasurer; Thomas B. Adams, was assistant treasurer and Edward Brush was secretary. The last two came from Meyer's company.[44]

At the moment of its inception, ASARCO controlled about two-thirds of the nation's smelting and refining capacity, but the corporations outside its orbit were as significant as those within. Hill and his associates tried to sell the Boston and Colorado firm to the underwriting syndicate, but the negotiations fell through, perhaps because the Argo works had never processed silver-lead ores. Nor did ASARCO acquire the three smelting companies on the Pacific coast—the Selby enterprise near San Francisco and the Tacoma and Puget Sound firms in Washington. Still another independent was the Compania

Metallurgica Mexicana, which failed to enter the combination even though Towne, Meyer, and Witherell served as directors of both concerns. Yet, important as these corporations were, by far the most significant lying beyond the control of ASARCO was M. Guggenheim's Sons. All these enterprises, but particularly M. Guggenheim's Sons, were soon to play important roles in the destiny of the American Smelting and Refining Company.

Chapter 9

Reduction in the
Age of ASARCO

ONCE THEY TOOK CONTROL OF ASARCO, EDWARD W. NASH AND his colleagues turned their attention to centralizing operations and coordinating their far-flung network of smelters and refineries. To facilitate the work, they delegated responsibility to various groups. August R. Meyer headed the new ore-purchasing committee, James B. Grant the operating committee, and so on.[1]

Rather than move to New York, where the firm established its corporate offices, Grant began rationalizing operations from his old headquarters in Denver. And perhaps this was a logical choice, since about half of ASARCO's plants were in Colorado. Once in charge, Grant and his lieutenants outlined uniform assaying procedures, standardized ore-purchasing policies, and apportioned shipments to various smelters. Soon it became common for Grant's traffic directors to have large shippers like the Ibex Mining Company of Leadville send consignments to four or five different plants. Such practices enabled ASARCO to carry lower stocks of mineral, yet gave metallurgists greater flexibility in preparing reduction charges. The arrangement also freed working capital that the enterprise needed. Yet Grant and his associates had barely commenced work when they found themselves confronted by a serious labor dispute, one that was to have unforeseen consequences for ASARCO.[2]

Shortly before the company's formation, the Colorado General Assembly had passed a bill giving workmen an eight-hour day, and Governor Charles S. Thomas had signed the measure into law, taking effect on June 15, 1899. The statute was in line with the national objectives of the labor movement, but the business community opposed it with the claim that it was unconstitutional for a state government to regulate hours of work. Yet this was little more than a smokescreen to hide the real issue. These men were simply opposed to organized labor, which had pushed the bill, and their opposition took on an emotional component because the law drew ardent support from the Western Federation of Miners, a union known throughout the West for its militant postures and quasi-socialist views.

The controversy surrounding the eight-hour statute set the stage for a major struggle between ASARCO and its employees. Like most business leaders in the state, particularly those in the minerals industry, Grant and his colleagues looked askance at both the eight hour law and organized labor. In particular, they opposed the Western Federation of Miners, whose subdivision, the Mill and Smelterworkers Union, was recruiting members in the company's plants.[3]

Somewhat surprisingly, the conflict began in the San Juan two weeks before the eight-hour bill became law. On the morning of June 1, Franklin Guiterman posted a notice outlining the firm's plan to pay workers on an hourly basis. The company would not require anyone to work more than the eight hours specified by the new statute, but each man would have that option. Once he had posted the notice, Guiterman left Durango on a business trip, but he gave instructions that only he could countermand the order. That afternoon a committee from the smelter workers' local told Guiterman's lieutenants that the union would go on strike the next day if they did not remove the notice. Guiterman's subordinates, however, either would not or could not alter his instructions. With the situation at an impasse,"150 men, nearly the entire work force, walked off the job on June 2. Mines in the San Juan began to close.

During the next two weeks Grant tried to head off a general strike at the plants in Leadville, Denver, and Pueblo. He and his colleagues decided to offer their employees hourly pay plus a 10 percent increase in wages. This in effect would restore the pay scale prevailing in 1893—provided each man stayed on the job for twelve hours. The smelter workers rejected the offer. They wanted ASARCO to recognize the union, and they wanted to preserve the existing wage rates—but for eight hours' work. This was not unreasonable, for pay in the industry was low. Even so conservative a mining man as John F. Campion, no friend of labor, thought it would be impossible for a man to obtain the necessities of life if the enterprise reduced the wages of unskilled workers from the prevailing scale of $1.75 daily.[4]

Campion's views notwithstanding, the men of ASARCO refused to budge. They thought that any pay increase exceeding 10 percent would eliminate profits, and, in fairness to them, their enterprise was new, its future uncertain. But Grant and his associates were opposed in principle to unions, particularly so militant an organization as the Western Federation of Miners. For a whole week Grant encouraged discussions between his plant managers and the smelter workers, but the talks failed to break the impasse. When the eight-hour law took effect on June 15, a majority of the workmen at Leadville, Denver, and Pueblo took the furnaces out of blast, then walked out. The general strike was on.

The stoppage affected the entire region. The smelters announced they would purchase no more fuel and ore until the walkout came to an end, a pronouncement that prompted layoffs in the coalfields and hard-rock mines. One observer went so far as to predict that thirty thousand men would be out of work if the strike lasted more than thirty days. Union members elsewhere in Colorado set up picket lines to show support for the smelter workers, while other workmen struck here and there over the same issues.

But not all the smelters closed. In 1893 Nathaniel P. Hill and his associates had not reduced the wages of their employees, and now in 1899 they arranged a compromise on the eight-hour issue acceptable to their workmen. The Argo works remained in operation, its reverberatories running steadily throughout the summer. And in Pueblo Simon Guggenheim negotiated long hours with smelter workers at the Philadelphia plant as the fifteenth of June approached. Finally, on the evening of June 16 the two sides reached a compromise that called for increases in pay ranging from 25 to 40 percent, the highest raises going to the lowest-paid. Guggenheim telegraphed the details to the family offices in New York, then a tense period of waiting followed as his father and brothers considered the pact. After lengthy debate they decided to accept it. And so during the summer of 1899 the Guggenheims ran their plant at full capacity while the men of ASARCO fought what proved to be a long, hard strike.

Meanwhile, efforts developed to resolve the dispute between the "smelting trust" and its employees. When news of the Guggenheims' settlement appeared in the press, the smelter workers union offered to end the walkout if ASARCO would agree to a similar compromise. But Grant and his colleagues were unmoved by the suggestion. Then the State Federation of Labor offered to serve as mediator. In a letter to Grant the organization's executive board requested a conference with representatives of the smelting company not only on behalf of the strikers, but also in the interest of the railroad, mining, milling, and manufacturing industries who, said the board, found their prosperity menaced by the closing of ASARCO's plants. Grant replied that he was aware of the "gravity" of the situation, but he thought that the "economic condition of the mining and smelting industry" precluded any possibility of increasing wages beyond the current offer. He added that he was willing to discuss the question with his employees, but not with the federation, since it knew nothing of the reduction business.

Only two days after the strike began, Governor Thomas decided to intervene because the walkout posed a threat to the state's economy. On June 17 he appointed a citizens' committee to look into the dispute and recommend a solution. This group, which included former governor Alva Adams, met with representatives of both sides. Grant reiterated his original proposal of hourly pay with a 20 percent wage increase. The smelter workers countered with the Guggenheims' scale, which would have provided the strikers a raise greater than ASARCO's offer but lower than their original demand. The citizens' committee held several more meetings, but when neither side showed any willingness for further compromise the group dissolved, having achieved nothing.

As soon as it became obvious that this effort had failed—and it became obvious in a very short time—the union maneuvered to bring the State Board of Arbitration into the dispute. The General Assembly had created this agency in hope of preventing strikes and lockouts, and it consisted of three members appointed by Governor Thomas. On June 28, with the walkout nearly two weeks old, the smelter workers' local in Denver asked the board to intervene. The next day William F. Hynes, secretary of the organization, asked Grant to let the conflict be settled by binding arbitration. But Grant remained firm in his position. He replied that the company had no business relations with the union and that it would not deal with it. The enterprise intended "to exhaust all legitimate resources" in resolving the issues before it would resort to a compulsory settlement.

The Board of Arbitration then took more forceful action. It had the power to compel witnesses to attend its deliberations and give testimony. When ASARCO received a subpoena, Grant announced that the firm would comply, but he insisted that the enterprise would not be bound by the board's conclusions and would not recognize the union. The remarks of ASARCO's president, Edward W. Nash, were even more forceful. He was quoted as saying that the firm would rather close its doors than tolerate such an organization. He and his associates would run the business to suit themselves and would not accept the dictation of other parties.

The smelter workers adopted a more conciliatory attitude, for they had much to gain—and ASARCO much to lose—by the board's action. The board was at least moderately prolabor and, as in most arbitrations, was likely to choose a compromise position that would benefit the unionists. They immediately proposed to end the walkout if ASARCO would agree to be bound by the board's findings. But Grant and Nash rejected the offer. They realized that the committee was likely to find a middle ground that would mean an increase in wages beyond the standard offer and knew that the committee might also force them to recognize a subdivision of the Western Federation of Miners.

While maneuvering went on in Denver, the strike front began to crumble. The smelter workers at the Bi-Metallic plant in Leadville had not joined the union but had walked out in sympathy with their comrades elsewhere. In late June they decided to abandon the picket lines and return to work on the basis of ASARCO's standard offer, hourly pay and a 10 percent increase in wages for twelve hours' labor. By early July the smelting company had the first of its furnaces back in blast.

About two weeks later the unionists received a serious blow. On July 17, the supreme court of the state of Colorado declared the eight-hour law unconstitutional—just as the more conservative businessmen had contended from the outset. This decision cut the ground from under the men on the picket lines. The strike at the Arkansas Valley plant collapsed immediately, and by the end of the month its furnaces were back in blast. The Pueblo and Durango works followed quickly.

Yet the Denver smeltermen continued to hold out. In late July the Board of Arbitration met with both sides and submitted a report calling for a compromise. But the findings were not compulsory, and ASARCO, its position immeasurably strengthened by the supreme court's decision and the strike's collapse elsewhere, rejected the proposal. The unionists continued to man the picket lines around the Globe and Grant smelters, but with their resources nearly exhausted the smelter workers could not remain in strike much longer. Two more weeks passed. Then on August 13 the strikers agreed to accept the firm's offer and declared the walkout over. By the end of the month ASARCO had put its Globe and Grant smelters back in operation.

Nash and Grant and their associates had defeated the unionists, but it was a Pyrrhic victory. The strike had disrupted the firm's cash flow and eliminated about $380,000 in revenues, according to one estimate. Nash and his colleagues could ill afford so large a loss so soon after the company's formation. Equally bad, management had wasted its time and energy in combating a walkout instead of concentrating its efforts on reorganizing the firm's internal structure. In retrospect, however, it is hard to see how the men of ASARCO could have reacted differently given their views on the eight-hour law, their conviction that the enterprise would be doomed if they made concessions, and their unyielding opposition to the Western Federation of Miners. Nonetheless, the bitter strike of 1899 was only a harbinger of turbulent labor-management confrontations that would beleaguer the firm in the years to come.[5]

Once operations resumed, the leaders of ASARCO gave their full attention to forging a smooth-running industrial empire out of the disparate parts they controlled. This objective required the firm to close many plants, for the business had excess capacity, as the smeltermen had long claimed. Nash and his colleagues began dismantling seven complexes during the first eight months of the firm's life, and the wrecker's victims included the Hanauer and Ibex smelters in the Salt Lake Valley, two plants in the Chicago area, the Pittsburgh and El Carmen works, and the long-idle American smelter in Leadville. The next year Nash and his colleagues closed both the Union and the Bi-Metallic plants in order to consolidate their Leadville operations at the Arkansas Valley works. Nine plants gone in scarcely one year! Yet all was not destruction. The firm had Karl Eilers build a large, integrated smelter at Murray, Utah.[6]

ASARCO ended its turbulent first year with a profit of $3,500,000, a relatively small return on a nominal capitalization of $65,000,000. It enabled the company to pay three dividends amounting to $1,500,000 on the preferred stock, but the holders of common stock received nothing because the firm needed the rest for plant improvements and working capital. Yet even this proved inadequate. ASARCO needed new capital, a need that pushed Nash and his associates toward other mergers. On March 20, 1900, the directors authorized the officers to ascertain the terms and conditions by which the enterprise might acquire its two principal competitors, Robert S. Towne's Compania Metallurgica Mexicana and the firm of M. Guggenheim's Sons, both thought to be rich in cash.[7]

By 1900 Towne had converted "the metallurgical company" into the largest integrated mining and smelting firm in Mexico. Using capital from Boston, New York, and Europe, he had acquired mines, erected a huge smelter at San Luis Potosí, and developed several ancillary firms, most notably the Mexican Northern Railroad. Many thought that Towne's company had a strong cash position, but actually the collapse of silver had destroyed Towne's prospects, forcing him to sell bonds to pay the dividends required on the preferred and guaranteed stock. Many shareholders were disappointed that Towne had not sold his controlling interest to the syndicate that created ASARCO.

How much Nash and his associates knew about the internal affairs of the CMM is not certain , but they must have known a great deal, for Towne and August R. Meyer were directors of both firms. Nonetheless, James B. Grant visited the mines and works of the company and examined its books. Then he sent his report to Nash. As might have been expected, Grant advised against acquiring the enterprise. His line of reason never appeared in public, but it seems likely that he must have noted the large debt and other obligations. Towne's company would only add to ASARCO's burdens. And there the matter ended.[8]

Meanwhile, what about the Guggenheims? They had not sold the firm of M. Guggenheim's Sons to the syndicate that created the "smelting trust," although they had participated in the merger talks. During the negotiations Moore and Schley had offered the family about $11,000,000 in the combine's stock in exchange for a reduction empire estimated to be worth $5,000,000 to $8,000,000. It has always been said that Meyer Guggenheim opposed the deal because he did not want to see the plants pass into the hands of an enterprise that he and his sons did not control. This may well be true, but one wonders. The Guggenheims had much more at stake in other quarters.[9]

Even as ASARCO came into being, the family was engaged in forming a new corporation. In February 1899 they obtained the support of a group of New York and London financiers that

Simon Guggenheim about 1908, while a United States senator from Colorado. Like Nathaniel P. Hill and James B. Grant, he converted his success in business into success in politics. Colorado Historical Society, Denver.

included William C. and Harry Payne Whitney along with Sir Ernest Cassel, and together with them they launched the Guggenheim Exploration Company, sometimes known as

Guggenex. The family acquired only about 17 percent of the outstanding shares in this venture, but Daniel Guggenheim and his brothers directed nearly all operations. Over the next few years this firm acquired many mining properties in the United States and Mexico, and most of the ores produced went to the Guggenheim smelters.

The family also benefited from the strike that idled ASARCO's plants in the summer of 1899. When the passage of the eight-hour law created time and wage questions, Simon Guggenheim negotiated a settlement acceptable to labor and management. This was fortunate because it enabled the Philadelphia smelter at Pueblo to remain in blast while its rivals in Denver, Pueblo, and Leadville shut down. As a result, mining companies in Colorado overwhelmed the Philadelphia works, leaving Simon Guggenheim and his chief lieutenant Edgar L. Newhouse with their pick of the highest grades of ore at the most favorable smelting rates. By the end of this turbulent year the Guggenheims' empire in metal stood in excellent financial condition, and the family had ready access to the capital markets of New York.[10]

It was this strong position that prompted talks between the Guggenheims and ASARCO. Negotiations began in the summer of 1900, scarcely a year after the original merger, and in the fall Nash and his associates offered the family a lucrative package. They would increase ASARCO's nominal capitalization from $65,000,000 to $100,000,000 then give the Guggenheims $35,000,000 worth of this stock—half in common, half in preferred—in return for all the plants of M. Guggenheim's Sons, "good will," and $7,500,000 in cash or its equivalent. Yet this offer was not quite so breathtaking in reality as it appeared on paper, and the Guggenheims knew it. The market price of ASARCO's stock was far below the nominal par value of $100 per share, and the family had to provide $7,500,000 in cash. Still, the "smelting trust" had tendered far more than Moore and Schley in 1899. But the Guggenheims were in no hurry. They demurred, and talks were broken off until after the national elections that November.

With President William McKinley and his Republicans safely

reelected, the Guggenheims and the management of the American Smelting and Refining Company resumed negotiations. In January 1901 they reached a tentative agreement calling for ASARCO to increase its capitalization from $65,000,000 to $100,000,000—half in preferred stock and half in common. Of this the Guggenheims would receive the $35,000,000 in new securities plus the $10,000,000 in stock remaining in the treasury from the time of the original merger two years before. In return ASARCO would obtain the four plants belonging to the firm of M. Guggenheim's Sons plus $6,000,000 in cash and another sum equal to two-thirds of ASARCO's working capital as of January 1, 1901, a figure agreed upon as $6,067,000. Yet the Guggenheims did not have $12,067,000; so in the settlement Henry H. Rogers offered to provide them with $6,000,000 in cash in return for the $10,000,000 worth of stock in ASARCO's treasury. The merger now lay in the offing. The first public indication of what was about to happen appeared on January 19, 1901, when the directors of ASARCO called for a special meeting of the stockholders to be held on February 16 to consider the acquisition of M. Guggenheim's Sons.

Now came the complications. After the union, Nash and his colleagues intended to market copper through the Guggenheims' selling agency, which ASARCO would acquire in the merger. This was a logical decision but it was not acceptable to Rogers and Lewisohn. They wanted to continue handling the product through their firm, the United Metals Selling Company—which was the reason they had financed the original combination. When the Guggenheims and ASARCO's management, now acting in concert, insisted upon using the family agency, Rogers and Lewisohn resigned from the board of directors. On February 15, the day before the stockholders were to vote, Rogers and Lewisohn had their lieutenants obtain a temporary injunction forbidding the merger on the grounds that it would produce a monopoly in restraint of trade. A day later the judge modified his ruling to permit the shareholders to vote on the proposed consolidation on the condition that should it be approved—and it was—the union should not go into effect until the New Jersey courts should try the case.

During the next two months the adversaries maneuvered on two levels. They contested the injunction publicly in the New Jersey courts, although the question of monopoly in no way resembled the real issue. At the same time Daniel Guggenheim, Nash, and their colleagues quietly negotiated with Rogers and Lewisohn. In the midst of these deliberations a lower court denied the injunction on March 2, but Rogers and Lewisohn had their lieutenants appeal the decision.

The climax to the struggle came on the afternoon of April 18. Guggenheim, Rogers, and their associates resumed the clandestine talks at the law offices of Samuel Untermeyer, who handled much of the Guggenheims' legal work. Both sides were approaching an agreement, and discussions continued into the evening at Delmonico's, a posh restaurant in New York. There the two groups resolved their differences. The men of ASARCO agreed to grant the United Metals company a five-year contract for the sale of copper. This arrangement would be more expensive than selling through the Guggenhims' agency, but at least the smelting enterprise would receive a greater return than before. Rogers received from the Guggenheims $10,000,000 in ASARCO's stock and provided them with the $6,000,000 they needed to complete their obligations. Rogers and Lewisohn then withdrew the suit, but by then it no longer mattered, for the New Jersey courts had denied the appeal earlier in the day. On the night of April 18 ASARCO's representatives filed the appropriate papers in the state capitol at Trenton.[11]

Once the stock transfers had been completed, the Guggenheims took control of the enterprise. Daniel was elected chairman of the board of directors and chairman of the executive committee. He would be chief architect of the firm's destiny for the next two decades. His brothers Simon, Morris, Isaac, and Solomon also took seats in the directorate. Nash remained president, but overnight he found his role diminished. Isaac replaced Winthrop E. Dwight as treasurer, and Solomon superseded August R. Meyer as chief of the ore purchasing committee. Meyer retained his seat in the directorate, but as time passed he had less and less to do with entrepreneurial affairs. James B. Grant and Anton Eilers held their positions.[12]

Under the leadership of Daniel Guggenheim, the American Smelting and Refining Company continued the reorganization launched by Nash and his associates. To coordinate all operations in the Rocky Mountain region, Guggenheim had the firm establish a Western Executive Committee, initially headed by Grant. When he had a heart attack in 1902, the enterprise reformed this loose organization into individual departments. Franklin Guiterman emerged as general manager of the Colorado Department and became the firm's chief operative in the Rocky Mountains.[13]

During this time zinc assumed a far more important role than previously. For decades the smelters had penalized ores holding even small quantities of the metal because it interfered with the formation of silver-lead bullion. In the late nineties, however, the price of zinc had risen sharply because of an increase in demand and the efforts of producers in Missouri to restrict their output. The situation had prompted O. E. Jacobson & Company of New York to ship small consignments of ore from the high country via Galveston, Texas, to a smelter at Nurpelt, Belgium. As the experiment returned a profit, the Jacobson enterprise began shipping large quantities of mineral drawn from the shafts and stopes of the A. Y., Minnie, Maid of Erin, Iron Silver, and other mines at Leadville. This situation resembled the one prevailing in the minerals industry many years before, and of course it was not to last.[14]

The Guggenheims, who still owned the A. Y. and Minnie mines, saw the potential for profit in treating zinc ores in the United States. In June 1901, shortly after they merged their interests with the "smelting trust," the major figures in the combine formed a subsidiary corporation, the United States Zinc Company, setting the capitalization at $1,000,000. ASARCO acquired the controlling interest and later bought all the outstanding stock. Daniel Guggenheim was elected the first president but later gave way to Nash. Grant became agent in Colorado, but after his heart attack he turned his responsibilities over to Guiterman.[15]

It took Guggenheim, Nash, and Grant another two years to get their projected plant on stream. Knowing that ore supplies would come primarily from Leadville, they decided to erect the works outside Pueblo and integrate operations with their three smelters in town. Then for technology. Because no one employed by ASARCO had much, if any, experience with the methods to be used, Guggenheim and his associates looked to Europe for the metallurgists who would install what was known as the Belgian retort process. In the fall of 1901 Joseph Schulte and K. Suhlberg arrived in Pueblo to duplicate the technology used at Nurpelt, Belgium. Simon Guggenheim later claimed that the plant was "largely in the nature of an experiment"; but in fact the United States firm chose proved equipment, installed and run by experts.[16]

Not until June 8, 1903, did the company open its works at Blende, a new town that took its name from a characteristic zinc ore. The plant had a capacity of sixty tons daily and produced a compound known as spelter, which averaged about 88 percent zinc. This the plant shipped east for marketing. The residue consisted of another product about 10 percent lead and 3 to 5 percent zinc that was sent primarily to the Eilers smelter for further working.[17]

Once it was in operation, Nash and his associates kept the Blende plant running steadily. The zinc production of mines in the high country rose sharply during these years, and Leadville took the lead. Properties like the A. Y., Minnie, Ibex, and Yak put out new riches. Nash and his colleagues expanded their capacity to keep pace with production, but despite their efforts they could process only a small portion of the output because of strong rivals in Colorado, Kansas, and Missouri. Nonetheless, the production of spelter rose steadily until 1907, when a financial crisis precipitated a sharp decline in zinc prices and forced many shippers to close. Nash and his associates cut wages in an effort to maintain profit margins. Two years later this prompted a strike as the employees walked out in hope of regaining the old wage scale.

The United States Zinc Company provided a small but significant addition to ASARCO's business. The metal gave the parent firm another product to market, and the plant increased the lead supplies of the Pueblo smelters, which always needed the base metal to collect silver. Yet the works also augmented ASARCO's drive for integration, since the bulk of the ore supplies came from mines controlled by the Western Mining Company, another subsidiary of the "smelting trust." The activities of the United States enterprise represented another American effort to draw upon the technology of Europe—an old theme, to be sure, but one that was to continue well into the twentieth century.[18]

As the Blende plant came into being, the men of ASARCO continued to integrate operations by forming another enterprise—the Carbon Coal & Coke Company. The chief responsibility fell to Franklin Guiterman, then emerging as ASARCO's chief manager in Colorado. Under his direction Carbon Coal & Coke acquired a large mineral deposit in a place known as Reilly Canyon, about seven miles southwest of Trinidad in the Spanish Peaks region. Guiterman opened the mines, erected a washery, and built a coking plant that eventually consisted of 350 beehive ovens. To guarantee the works a steady supply of labor, Guiterman had the firm construct a company town known as Cokedale, which had 1,300 inhabitants by 1909.[19]

Meanwhile, the Guggenheims and their associates continued to streamline their main operations by phasing out obsolescent or unneeded smelters and refineries. In 1901 the enterprise closed the Great Falls, Mingo, and Argentine plants in the United States and the small Antofogasta works in Chile. The next year the firm shut down the Germania and Philadelphia smelters, although the latter continued working small quantities of copper minerals until 1905, three years after its last shipment of silver-lead bullion. Yet Karl Eilers finally opened a large, integrated smelter at Murray, Utah, and this plant took possession of the ore shipments that had previously gone to the old plants in the Salt Lake Valley.[20]

By the end of 1901, ASARCO virtually controlled the market for Colorado's production of smelting ores. The Argo works and a few other firms offered a measure of competition for certain classes of mineral, but these enterprises had relatively small capacities in comparison with their giant rival. ASARCO's managers regulated the output of the mines and sometimes increased reduction charges, which contributed to more friction between miners and smelters. Throughout the high country, many mineowners and managers deplored the company's preeminence and called for new smelter construction to combat the inequities of the "trust." Others wanted the federal government to "bust" the combine, which one man claims was "crushing the life out of the mining industry"—as he wrote President Theodore Roosevelt.[21]

Yet no one deplored ASARCO's position when the firm raised its ore prices. Early in 1903 the enterprise passed on to mining companies a general advance in the price of lead. This stimulated production from marginal properties at Leadville, Aspen, Silverton, and other camps and provided greater profits for the large shippers who were the mainstay of the corporation. But changes in the ore schedules were hardly altruistic. The enterprise controlled about 80 percent of American lead production, and fluctuations in the pay to mining companies tended to reflect the firm's desire to stabilize prices.[22]

But if mineowners were pleased by this turn of events, labor was not. Smelter workers still wanted higher pay, shorter hours, and better working conditions, and in 1903 these objectives precipitated another imbroglio between ASARCO and its employees. This time the trouble was confined to Denver. The preliminaries began on the first of May when the mill and smelter workers' union voted to demand an eight-hour day at the Globe and Grant works, but no one expected ASARCO to accede without a struggle. To increase their negotiating strength, union leaders spent the next six weeks recruiting new members and preparing for an almost inevitable strike. Then on June 17 a committee of the local politely opened the issue by sending a letter to Franklin Guiterman, now head of the Colorado department, headquartered in Denver. They requested that the firm grant all its employees an eight-hour day on the grounds that other smelter workers had it and that furnace work was both hazardous and unhealthful. They asked the firm to comply by July 1.

Guiterman was out of town on business when the letter arrived at his offices, and nearly two weeks passed before he replied. In a note to the local, he emphasized that ASARCO did not recognize a union—which was the standard position of the firm. Then he took up the question of the eight-hour day. He pointed out that three-quarters of Colorado's smelting ores came from just three camps—Leadville, Aspen, and Creede—but since 1900 their gross tonnage had fallen 20 percent, coupled with another 15 percent decline in silver content. Adding to these problems, said Guiterman, was a decrease in the value of silver itself. To operate under such conditions and prevent mines from closing, the management had found it necessary to lower reduction costs and smelting fees through close working and better mineral distribution. In light of this, he had to deny the union's request on the grounds that an eight-hour day would increase smelting costs, these would have to be passed along to mining companies, and thus many firms would have to cease production.

With Guiterman's denial, polite exchanges gave way to militant action, as both sides revealed their hard attitude. On the evening of July 3, union members gathered at the Elyria town hall and voted overwhelmingly to strike the Grant and Globe smelters. Shortly after the meeting, a small group of men, perhaps twelve to twenty, marched to the Grant works. There they induced two hundred workers on the night shift to join the walkout. In an orderly fashion the fires were drawn and the electric lights turned off, although it has always remained unclear whether someone used the master switch or simply cut the cable. What workers, if any, wanted to remain on the job had to leave because of darkness.

From the Grant works the strikers headed toward the Globe smelter, but now, according to published accounts, they degenerated into an uncontrolled mob. As they reached the Globe plant, they encountered a wire fence, but this proved a small deterrent. The unionists stormed the gates, broke through, and entered the smelter, where they convinced the night shift to join the walkout. Most workers did so because they belonged to the local, but several who wished to remain on the job were beaten up. As the men left the plant, the night superintendent and some clerical workers managed to draw metal from two furnaces, but that was all. Three others "froze up."

Violence played little part in the rest of the strike, but it had already destroyed the union's public image, costing the workers dearly. Throughout Colorado the press denounced such outbursts as intolerable and directed strong criticism at the local. Equally bad, the violence reinforced the smelter owners' view that war prevailed between the company and the union. There could be no compromise. Grant himself declared that he and his colleagues were "in the fight and will be there at the finish. What is the use of giving in?"

Both sides then maneuvered for advantage, but ASARCO had the upper hand. The walkout idled 775 men, a good portion of whom set up picket lines around the Grant and Globe plants. On July 7, four days after the strike began, Guiterman obtained an injunction that prohibited picketing by the Mill and Smelterworkers local as well as by the parent Western Federation of Miners, the American Labor Union, the Denver Trades and Labor Assembly, and the Colorado State Federation of Labor. This injunction also accused the union officers of conspiring to thwart operations of the smelters. Simultaneously, Denver's ultraconservative, antiunion Citizens' Protective Alliance offered to help prosecute "lawbreakers"—meaning strikers—and recruit nonunion workers.

Maneuvering extended far beyond Denver. Even though the firm had closed several of its Colorado smelters, it still had excess capacity. To outflank the strikers, traffic directors diverted ore shipments destined for Denver to plants in Leadville, Pueblo, and Durango. Once the unionists realized that the continued operation of these smelters jeopardized the strike, they dispatched representatives to persuade fellow workers to join the stoppage. But ASARCO's managers countered this move by convincing other employees to remain on the job and abide by the decision reached in Denver. Citizens' alliances also harassed union organizers and drove them out of town. The Western Federation of Miners managed to stage a sympathy strike in Cripple Creek, and two-thirds of the work force temporarily shut down the Durango smelter, but these events had little influence on the situation in Denver.

Guiterman put greater pressure on the strikers during the

tense summer months. Two weeks after the walkout began, he reopened the Globe works with twenty-five nonunion men protected by the Denver police. Later he tried to import strikebreakers, an effort that failed. Then came a draconian move. The firm announced that it had decided to close the Grant smelter permanently in conjunction with its desire to streamline operations and phase out obsolescent plants. This may have been true, but it struck a hard blow at the strikers, for in one swift stroke ASARCO had eliminated 475 jobs. The Globe works had only 300 positions.

Yet the strike continued. Summer gave way to fall, and fall dragged on into winter—a long time to be out of work—and time was on ASARCO's side. As the unionists exhausted their resources, some reapplied for jobs. By November Guiterman had four smelting units back in blast at the Globe plant. Operations soon returned to normal, and the walkout withered away. Only bitterness remained.[23]

By the end of 1903 the American Smelting and Refining Company had sharply altered the complexion of the reduction industry in Colorado. The firm had consolidated its operations at the Arkansas Valley, Eilers, Pueblo, Globe, and Durango works. Gone from the business in the span of four years were the Grant, Philadelphia, and Bi-Metallic plants that had been important in western ore reduction for nearly two decades. ASARCO had only two rivals of any consequence: the old Boston and Colorado Smelting Company and the new Ohio and Colorado Smelting Company.

Yet even now ASARCO's ore buyers could not find enough mineral to keep all the furnaces in blast. The firm offered to pass at least a portion of its lower reduction costs on to mining enterprises, but this failed to stimulate production. The state's output remained high, but the shipments of relatively high-grade ores able to pay the cost of transportation and reduction by the smelting process were on the decline. Newer technologies like ore flotation and cyanidation were coming into their own as the best ways for treating low-grade mineral. These trends became ever more pronounced in the next few years.[24]

Meanwhile, as the Colorado department consolidated operations, contended with labor unions, and tried to increase ore production, ASARCO received a threat to its prosperity, if not survival, from another quarter. Early in 1903 Charles Sweeny, a controversial promoter, mineowner, and entrepreneur in the Coeur d'Alene, conceived the idea of uniting the major mining companies of northern Idaho into a single enterprise. Such a combination had much to recommend it, for this hypothetical corporation might deal firmly with the truculent Western Federation of Miners and might negotiate a far more lucrative reduction contract with the American Smelting and Refining Company. Sweeny did not have enough money to undertake such a venture, but he thought the man who did was John D. Rockefeller, the best-known businessman in America. Not only did the oil magnate have what seemed to Sweeny unlimited financial resources, but, more important, everyone in the Northwest knew that Rockefeller had invested heavily in the minerals industry. Sweeny logically assumed that he might be interested in the scheme.[25]

Early in 1903 Sweeny traveled to New York. He had hopes of persuading Rockefeller to finance the combination, but he was never to see him. Instead, it took Sweeny about two weeks to obtain an appointment with Frederick T. Gates, Rockefeller's lieutenant in charge of all investments in the Pacific Northwest; and it was to Gates, a Baptist minister turned businessman, that Sweeny outlined his plan. Gates found the scheme attractive, and so did John D. Rockefeller, Jr., who was taking over many of his father's business interests. After talks with the persuasive Sweeny, they recommended that Rockefeller finance the combination.

But John D. Rockefeller was opposed, and for good reason. In the early 1890s the investment banking house of Colby, Hoyt & Company had persuaded him to purchase large blocks of stock in several mining companies owning claims at Monte Cristo, Washington. To integrate operations, Rockefeller had financed the construction of a concentrator in the mining town, a railroad from there to tidewater, and a smelter outside the city of Everett. Unfortunately, the mining engineer who assessed the Monte Cristo properties had made an egregious error. The veins pinched out a short distance below the surface, and the mines proved nearly worthless. Nothing except red ink flowed into the Rockefeller offices in New York. By the time Sweeny appeared in 1903, Rockefeller had written off more than $850,000 in losses. He was understandably reluctant when his son and Gates relayed the details of Sweeny's plan, which called for the investment of several million dollars.[26]

Despite this reluctance, Gates and the younger Rockefeller were convinced that Sweeny's design would solve the problems in Washington. The Coeur d'Alene mines were great producers, their ores might make the Everett smelter profitable for the first time, and at the least the proposed corporation would take over the smelter, the concentrator, and the dreadful mines at Monte Cristo. At last Gates and the younger Rockefeller persuaded the oil magnate to finance the combination, but only on condition that other prominent businessmen would invest as well, the most prominent of these being George J. Gould, who controlled the Rio Grande Railroad and who would be expected to give the new enterprise favorable freight rates. Gould consented.[27]

Once the senior Rockefeller gave his reluctant assent, events moved forward. Supplementing his own network of connections with capital from Rockefeller, Sweeny purchased options on the Standard, Mammoth, and Empire State–Idaho mining companies. Simultaneously, Gates and the younger Rockefeller organized the Federal Mining and Smelting Company. On July 24, 1903, this firm obtained a perpetual charter from the state of Delaware. Preparations continued over the summer, then in September Federal took possession of every important mine in the Coeur d'Alene except the famous Bunker Hill & Sullivan. Sweeny emerged as president of the combine, but the real control lay in the hands of the Rockefeller group.

Daniel Guggenheim and his associates at ASARCO viewed the formation of Federal as ominous. They realized that by controlling more than half the known reserves of Idaho, Federal could demand a far more lucrative smelting contract than could its predecessors. Yet this was the least of their worries. Far more unnerving was the sudden appearance of rumors that Sweeny and the Rockefellers were about to enter the reduction industry on a grand scale. This was a virtual death threat to ASARCO, because its plants in Colorado, Utah, and other states depended to a great extent upon the Coeur d'Alene for lead essential in smelting dry ores. Federal's smelter at Everett was a rival of little consequence—small, isolated, undercapitalized—but if Sweeny and his financial supporters integrated their operations forward into ore reduction, as was suggested by the firm's name, ASARCO might well be doomed. And Nash, Grant, Eilers, and the Guggenheims all knew it.[28]

In this portentous situation Daniel Guggenheim and his associates had to deal with the Rockefeller group—or so they thought. During the summer Guggenheim and other corporate officials had "accidentally" met Sweeny on the streets of New York to sound out his intentions. Yet these men were no fools. They perceived that the real power lay in the hands of Gates and the younger Rockefeller. Daniel Guggenheim arranged a series of appointments with Rockefeller's son. The truth was that John D. Rockefeller had no intention of entering the reduction industry; in fact, he had consented to support Federal largely to create a vehicle with which to extricate himself from a business in which he had lost hundreds of thousands of dollars. But the men of ASARCO could not know this. John D. Rockefeller, Jr., drove a hard bargain. He and Guggenheim negotiated a five-year contract by which ASARCO granted Federal a lucrative smelting rate, placed no restrictions on production, routed the ores over railroads controlled by Gould, and leased the Everett works, which the "trust" soon purchased outright under the terms of the agreement. In return, the Guggenheims bound themselves not to purchase any mines in the Coeur d'Alene—meaning the Bunker Hill & Sullivan—for another two years.[29]

This arrangement proved satisfactory for a time. Sweeny shipped huge tonnages of mineral, which provided ASARCO's smelters in Colorado, Utah, and elsewhere with the lead essential in working dry ores. Yet the situation continued to worry ASARCO because the Rockefellers might yet take Federal into the reduction industry. Rumors circulated to this effect, and Sweeny talked openly about plans to do so. Scarcely anyone knew that the Rockefellers had no intention of pursuing such a project.

In light of the unconfirmed reports, Daniel Guggenheim once again initiated talks with John D. Rockefeller, Jr., this time with an eye to acquiring control of Federal. As the negotiations proceeded early in 1905, both parties began buying the company's common shares, which had been listed on the New York Stock Exchange. This competition, coupled with a general uptrend in the market, drove the price of Federal's common stock to new heights, but in the quest for control the Rockefellers had the advantage. By February 1905 they garnered an absolute majority. Then came the climax to the struggle. In a final round of talks with John D. Rockefeller, Jr., Daniel Guggenheim agreed to pay $120 a share, or about $3,300,000, for the nearly 27,500 shares held by the Rockefeller group. Yet Guggenheim did not have this much cash on hand; so in the final transaction the oil magnate decided to accept $1,300,000 in cash and lend the smelterman the remainder. On March 16, 1905, Charles Sweeny delivered 28,105 shares of Federal's common stock to Guggenheim in return for $3,372,600, of which Rockefeller received $2,200,600.[30]

ASARCO, however, did not take direct control of Federal. A short time before Daniel Guggenheim consummated the agreement with the Rockefellers, he and his associates organized a subsidiary corporation known as the American Smelters Exploration Company, which had a life span of only three weeks before it was reorganized as the American Smelters Securities Company. This enterprise took possession of Federal. Sweeny continued as president because the Guggenheims thought he was a good mining man, but as before the real control lay in New York—only now it resided in ASARCO's offices at 120 Broadway.

Once they took possession of Federal, the Guggenheims moved to assure ASARCO of complete supremacy in the Coeur d'Alene. This meant controlling the production of the Bunker Hill & Sullivan Company, the only major firm in northern Idaho beyond Federal's domain. After winding up his talks with the

Rockefellers, Daniel Guggenheim turned his skills in negotiation here, and soon ASARCO signed a twenty-five-year contract with the Bunker Hill & Sullivan, thus heading off potential competitors and insuring its smelters of adequate supplies of high-grade lead ores for years to come.[31]

Having secured their northern flank, Guggenheim and his associates used the Securities Company to extend ASARCO's control over the mining and smelting industries. Even as Daniel Guggenheim negotiated for the acquisition of Federal, he and his colleagues employed Bernard Baruch, then known as a shrewd Wall Street investor, to purchase the controlling interest in two other reduction firms, the Tacoma Smelting and Refining Company of Washington and the Selby Smelting and Lead Company of California. Not only did the Securities enterprise take over these corporations, it also purchased the majority interest in the Guggenheim Exploration Company, whose subsidiary, the Western Mining Company, controlled the Ute, Ulay, Auric, and Silver King mines in the San Juan country, the A. Y., Minnie, Adams, Maid of Erin, and Wolftone properties in Leadville, and even more valuable ore deposits in Mexico.[32]

During these turbulent years the old Arkansas Valley works deep in the Rocky Mountains, ten thousand feet above sea level, emerged as ASARCO's most important smelter in Colorado. Even before the Guggenheims brought their capital and expertise into the firm, Nash and the original management had increased the plant's ability to process the ever-lower grades of mineral coming from mines in the high country. Late in 1900 the company appropriated a sum estimated at $325,000 to install additional roasting and smelting units. The task of making the improvements fell to Julius Rodman, who replaced Joseph H. Weddle as manager about this time. When completed, the new furnaces gave the plant a capacity of about one thousand tons daily.[33]

Once the Guggenheims entered ASARCO, they kept the AV works running at full capacity, although there were changes in personnel and practices. Rodman's tenure at the smelter was short. When he left to assume a position with the Western Mining Company, the firm replaced him with William B.

McDonald. Like his predecessor, McDonald proved resourceful in operating the plant, but like other smeltermen past and future he found himself unable to overcome dissatisfaction in the mining community, this time created when higher levels of management notified shippers in Leadville that in the future the smelter assays would prevail on all ore consignments unless there were huge discrepancies in determinations. In this case McDonald would send ore samples to Denver for the decision of an umpire. Some producers like the important Ibex Mining Company thought this arbitrary—a consequence of monopoly —and they ceased operation in protest. One mineowner complained that ASARCO's new system nearly always meant a judgment in the smelter's favor. Such protestations, however, were always temporary—no profits came from shuttered mines—and the AV works continued its steady reduction of a thousand tons daily despite the grumblings of the mining industry.[34]

Like these shutdowns, the bitter strike at the Denver smelters in 1903 had little real effect on McDonald's operations. Union representatives arrived in Leadville to persuade workers at the AV plant to join the walkout, but those efforts failed. McDonald kept the works operating at capacity throughout the dispute. Guiterman and the traffic managers rerouted some ore shipments to Leadville, and after a time this did create some aberration in McDonald's plans. By September, sulfide shipments had grown so heavy that ore cars choked the plant's railroad facilities, forcing McDonald to decline further consignments, which in turn compelled several mines to close.[35]

Even after smelting operations returned to normal, McDonald still had problems with excess sulfides. During the winter of 1904, he had to slow operations when silicious ores sent to be mixed with sulfides happened to freeze aboard railroad cars because of a labor scarcity that some attributed to the firm's low wages. A few months later an accident at the plant's power source compelled McDonald to curtail work at the sulfide mill. Soon the ore bins overflowed with fifteen thousand tons of mineral. McDonald asked the Ibex, Yak, Iron Silver, and other large producers to restrict their output, but they kept right on shipping according to contract. Then Guiterman took to his pen. He pointed out to one mineowner—John Campion of the Ibex— that such huge consignments were causing the AV works to incur "extraordinary expenses" that narrowed profit margins. Would the mine restrict its output? But Guiterman's pleas aroused no more sympathy than McDonald's. Men like Campion enjoyed ASARCO's plight and went right on shipping. To compound the problem, McDonald had to contend with a strike by wheelers and weighers who wanted an eight-hour day, but the walkout ended quickly with the twelve-hour shift intact.[36]

Soon the Guggenheims and their associates decided to increase the capacity of the AV works, a decision prompted largely by the rising production of low-grade sulfide ores in Leadville. In the fall of 1905 the firm appropriated enough capital for McDonald to erect two more blast furnaces and enlarge several others. Then came new technology. During the winter the men of ASARCO installed the Huntington-Heberlein process, another European development to which the enterprise had purchased the exclusive patent rights in the United States and Mexico. This method involved a form of blast roasting that lowered the cost of preparing sulfides for smelting. To make room for the new system, McDonald had to dismantle several old roasters. Output fell for a time, but by the spring of 1906 the works had a capacity of 1,400 tons daily, or about 500,000 tons yearly.[37]

These additions notwithstanding, McDonald still found himself deluged by sulfides. In spite of declining production here and there, Leadville and other mining camps sent huge tonnages to the Arkansas Valley works, and McDonald could never find enough silicious ores to mix with them in preparing furnaces charges. In May 1906 many employees left to find more lucrative work in Utah and Idaho, and ores piled up more rapidly than McDonald could have them smelted. This prompted Guiterman to write another round of letters asking ASARCO's "friends, the large shippers" to restrict their output because of the "frightful accumulation" of sulfides. Adding to this, he said, the railroads were hounding the firm to unload its ore cars. Yet the "friends" kept shipping according to contract, and the situation grew worse. By July nearly one-third of the work force had departed, and McDonald had to take several furnaces out of blast and run at three-quarters capacity.[38]

This situation continued into 1907. In January Guiterman reluctantly diverted some shipments to the Globe smelter in Denver, but the ores were so low in grade that they barely paid the freight tariff. This effort notwithstanding, McDonald still found himself unable to process the huge quantities of mineral shipped to the AV works for reduction. As a result the railroads charged the company $14,000 in penalty fees for failing to unload ore cars. When the difficulty persisted, Guiterman cut off the small shippers entirely, and several mines had to close because they had no other market than the Arkansas Valley smelter.[39]

Yet the problem of excess ores ended later that year. A new financial crisis that gripped the nation contributed to sharp declines in the prices of silver, lead, copper, and zinc. ASARCO met the situation by reducing its own returns on mineral, and many producers had to shut down because they could not operate profitably. Even when metal values rose two years later in 1909 and 1910, many potential shippers remained closed, particularly the smaller outfits that could not benefit from economies of scale in ore production. Leadville was at last on the wane as a mining camp, and after 1907 the Arkansas Valley smelter rarely worked at full capacity.[40]

The Globe plant, meanwhile, pursued a somewhat different course. After the long strike of 1903 it resumed operations as ASARCO's only smelter in the city of Denver. As before, it competed with the Argo smelter in the Clear Creek ore markets, drew other mineral from the central Rockies, tapped Cripple Creek for silicious ores, and reached across the continent for its remaining materials. And as before, it got its essential supplies of lead from mines in the Coeur d'Alene, a situation made easier once ASARCO took control of the Federal Mining and Smelting Company. By 1906 the Globe works, now under the management of Frederick Roeser, turned out 1,800 tons of bullion and another 200 tons of lead-copper matte monthly. And, what was more, the plant had the lowest smelting costs of any in

ASARCO's empire. These expenses became even less when the firm installed the Huntington-Heberlein process a few years later.

Yet, if smelting operations remained essentially the same, refining changed to a remarkable degree. The Globe had its own separating plant, the one built by Dennis Sheedy and Malvern W. Iles back in the early 1890s, but in the reorganization that followed the creation of ASARCO, Grant and his associates decided to phase out the refinery and send the bullion to Omaha for further working. The buildings remained, however, and after Guiterman became head of the Colorado department, he had new equipment installed so that the works could recover cadmium, thallium, and indium compounds captured in the bag houses used in all the firm's smelters.[41]

After 1907 the Guggenheims, Guiterman, and the other managers of ASARCO found themselves in an altered position vis-à-vis their Colorado smelters. The gross ore production of the high country remained steady, but the quantity of silver and lead in every ton declined sharply, as did the market value of the two metals. Over the next few years mining companies like the Ibex and Yak increased their output of complex zinc ores, keeping that industry alive and masking the decline of the more traditional elements. To process blende and other forms of zinc-bearing rock, rival enterprises erected plants in Oklahoma, Kansas, Texas, Illinois, and other states. All these works shipped large amounts of silver-copper-lead compounds back to ASARCO's smelters in Colorado for reduction, but the altered supplies of ore and new technologies for handling low-grade minerals changed the role of the smelting industry.[42]

In the years that followed the recession of 1907 and 1908, ASARCO's plants in Colorado rarely operated at capacity. The Globe smelter used about three or four of its seven blast furnaces, and the company lowered its wage scale in an effort to maintain profit margins. This led to short-lived strikes in 1910 and 1913 as the employees tried unsuccessfully to regain what they had lost. In Pueblo the enterprise closed the old Eilers plant in 1908, twenty-five years after the famous smelterman and his colleagues had broken ground. This left only the old works built by Mather and Geist in Pueblo, but in spite of the sharply decreased capacity the smelter rarely employed more than three or four of its smelting units. The Durango plant cut back to one furnace. The Arkansas Valley works maintained the largest capacity of any Colorado smelter, but rarely did the plant have more than five units in blast—half of those available. Even the high metals prices created by World War I failed to resuscitate the industry. Other firms with different technology received what benefits the conflict created.[43]

The Guggenheims and their associates could do little to prevent the collapse of ASARCO's ore markets. In July 1908 Guiterman wrote one mineowner that "for the first time in the smelting business I feel myself kind of groping in the dark." A year and a half later George E. Collins, another mining man, observed that one looked in vain for signs of prosperity, present or prospective, in the industry. He knew that ASARCO and its few remaining rivals needed new mining districts to provide additional supplies of ore, but he saw little chance of this happening. Prospecting was not energetic.[44]

As Denver saw its smelting industry collapse, the Chamber of Commerce appointed a special committee headed by Guiterman to investigate the decline and recommend measures to reverse the trend. In November 1910 he filed his report—not a propitious one. Not surprisingly, he reviewed the steady decreases in milling and smelting charges made over the years. Then he concluded with "disheartening disappointment" that lower reduction costs had not stemmed the decline of the mining business. Prospecting had nearly ceased, and, what was worse, the ore bodies that had sustained the minerals industry for decades were now virtually exhausted.[45]

By the onset of World War I nearly all the men who had guided the industry had passed from the scene. Malvern W. Iles had died in 1900; Edwin Harrison in 1902; August R. Meyer, Edward W. Nash, and Meyer Guggenheim in 1905; James B. Grant in 1911; and Benjamin Guggenheim in 1912. Edward R. Holden and Alfred W. Geist had vanished. Anton Eilers and

Franz Fohr had retired in the East, August H. Raht in San Francisco, and Otto H. Hahn in Germany, but they would all die before the guns of World War I grew silent. Only Daniel Guggenheim and his surviving brothers, members of a younger generation, remained. They controlled the remnant of ASARCO's Colorado empire from their offices in New York.

Chapter 10

"Groping in the Dark"

FRANKLIN GUITERMAN MAY HAVE VIEWED THE SMELTING industry with "disheartening disappointment" when he surveyed the field in 1910, but the same sentiments must have been felt in other quarters well before then. The early twentieth century brought even more challenging times to an already hard-pressed business. Yet the few remaining independents kept on working at least for a time, while new entrepreneurs still saw—or thought they saw—potential profits in reducing the lower and lower grades of ore coming from the high country mines.

In the early years of the century Crawford Hill became the chief architect in shaping the destiny of the Boston and Colorado Smelting Company, the state's oldest reduction firm. He had assumed an ever larger role in decision-making during his father's last illness, but he had never held an official position in the corporate structure. Once he took over full responsibility for western operations, he immediately sought to clarify his role in management. In letters to company officers in Boston he suggested that the board appoint him resident director, partly on the grounds that the firm was a Colorado corporation even though most of its stockholders lived in New England. This recommendation met with a favorable response, and the board named him to the position, much to Hill's satisfaction.[1]

Then came another question. Sometime during the nineties Nathaniel P. Hill had lent the corporation $275,000 for improvements to the Argo works. He had not insisted upon a formal schedule for repayment of principal and interest, largely because he had known the officers and directors for many years. Convinced of their friendship and integrity, he had been content to accept checks in varying amounts from time to time.

This arrangement, however, was not acceptable to Crawford Hill. He saw the president, Costello Converse, and other corporate officials in Boston primarily as business associates, not as friends of long standing. Thus he wanted the enterprise to establish a specific timetable by which it would repay this large debt to the family, and to this end he wrote George D. Edmands, the treasurer, that the Hills wanted quarterly interest paid on the outstanding notes. The question of money proved harder to resolve than the question of resident director, but after a year of proposals and counterproposals the Boston and Colorado firm agreed to exchange the Hills' outstanding notes for a new issue that carried annual interest of 4 percent paid quarterly.[2]

The death of Nathaniel P. Hill was a significant hour in the venture's evolution, but another of equal importance came in July 1901 when Richard Pearce resigned as manager of the Argo works. After nearly thirty years in the company's service this had been a hard choice, and Crawford Hill found the Cornishman "very unsettled" over his decision to terminate so long an association. Yet Pearce was now in his sixty-third year, the prospects of the firm appeared cloudy, and he may well have seen 1901 as a good time to make the inevitable severance.

Yet if Pearce saw this as a good time for him to retire, company officials in Boston did not. They delayed action on his resignation for several months. The reason was simple. Troubled over the firm's declining prosperity, Converse and others in the eastern management were beginning to question the integrity of the western staff. Pearce's retirement looked suspicious. In November, after an unseemly period of inaction, Hill finally urged Converse to have the board of directors adopt "very strong resolutions of deep regret" over the Cornishman's departure. Prodding brought results. The directors finally passed a resolution expressing their gratitude to Pearce, while both Converse and Moses P. White, the corporate secretary, wrote personal letters to the retiring manager. Whether Pearce ever learned about the flurry of suspicion is a matter of conjecture, but it was unfortunate that such suggestions marred the exit of a man who had long played so large a role in the firm's success.[3]

Despite his long years of residence in the United States, Pearce had never acquired American citizenship, and early in 1902 he returned to Great Britain. He intended to spend his remaining years in retirement, but his reputation in metallurgy brought an end to these plans. Soon after his return a smelting enterprise lured him into managing a plant in Liverpool. He also received an offer to stand for Parliament, although this opportunity he declined. Pearce worked several more years in Liverpool, then retired and moved to London. Here he lived out his last years, dying in 1927 a few months before his eighty-eighth birthday.[4]

In the meantime, as Pearce concluded his association with the Boston and Colorado Company, Hill perceived that the firm had entered a new economic environment. ASARCO, of course, was a strong competitor, stronger than anyone the enterprise had yet encountered, but this was only one dimension to the new scenario. What held ever greater portent was the steadily eroding production of mines on the forks of Clear Creek, the smelter's chief market. Hill realized that the decline foreshadowed an end to the prosperity of the firm. In this light he and the eastern management concluded that it might be best to sell the Argo works before profits turned into losses.

This decision prompted talks with the American Smelting and Refining Company. In the summer of 1901 Converse and his eastern associates approached ASARCO and negotiations ensued, but the two sides failed to reach agreement, if indeed they ever bargained seriously. Hope for a sale appeared gone, but in November, months after talks in the East had come to a fruitless end, Hill reported that the Guggenheims' manager in Denver had mentioned informally that he wished to inspect the Argo works and examine the books. Hill did not name the man, but he was probably either James B. Grant or Franklin Guiterman. Hill was rather surprised. He finally resolved to give this person a tour of the smelter and provide him with data on reduction costs, dividends, and ores on hand, but he would not divulge any details of the processes. Such plans, however, proved unnecessary, for the Guggenheims never sent a representative to Argo. By January Hill had concluded that it would be unwise to approach ASARCO again.[5]

While all this was going on, Harold V. Pearce took over as manager of the Argo smelter. He was the natural choice, since he had worked in the plant for some years and knew every aspect of its operation. Pearce assumed his position in July 1901, but Hill could not persuade the eastern management to give the new manager an official appointment until November, probably because of the negotiations with ASARCO. Once the designation came through, Hill asked Converse to give Pearce an annual salary of $5,000 as well as 5 percent of the profits.[6]

Yet if Pearce expected to supplement his pay from profit sharing he was to be disappointed, for the firm's economic position deteriorated in the fall of 1901. In October Hill warned officials in Boston that profits were dependent upon stability in the price of copper, and he feared that the metal's value might decline as much as four cents a pound. In this case the enterprise stood to lose about $88,000. Hill was apparently aware of an impending struggle between the independent producers and the Amalgamated Copper Company, which was about to make an effort to seize control of the market by selling its huge stocks to depress the price and force other enterprises to come to terms. Hill thought that any fall in prices would be temporary and that once Amalgamated and its rivals resolved the struggle they would curtail output and raise the value even above previous levels.

While Hill's views offered solace for the long run, they did little to bring relief during the winter. As the value of copper declined, producers of ore and matte held back shipments to Denver in hope of realizing higher returns later. Fewer consignments, however, created problems at the smelter, for the metal was essential in the Argo process. With supplies of copper running short, Pearce and Hill decided to recycle what stocks they had on hand so that they could continue working gold and silver ores. But this decision coupled with the low price brought on the loss Hill had predicted.[7]

This combination of events—Pearce's unexpected resignation, the failure of talks with ASARCO, and Hill's predictions of a loss—created suspicion in Boston. In a letter to Hill, Converse questioned the integrity of the western management and suggested an examination of the books to determine if there had been any chicanery. Hill took exception to this. He declared that there was "no question for any concern about the honesty" of the Argo personnel. His uncle, Jesse D. Hale, controlled the local checking account and the Pearces handled all the company drafts. Hill's assurances caused the problem to pass over for the time, but the suggestion of corruption surfaced again in a later year.[8]

Hill had forestalled the question of integrity, but he had no such luck in handling the scarcity of copper-bearing ores; they remained in short supply even though the price of copper rose in 1902, just as he had predicted. The sale of the Butte enterprise had already eliminated Montana as a dependable source of mineral, and smelter construction in other towns had cut off additional markets. And now ASARCO electrolytically refined the matte produced in its own blast furnaces. This not only removed another source of base metal but also did away with still another lucrative portion of Argo's business. In reflecting on the problem in later years, Harold V. Pearce said that it was not until the abundant supplies of copper vanished early in the century that the company came to realize just how valuable the metal was in smelting ores.[9]

In addition to the dearth of copper, Hill and Pearce found themselves confronted with shortages of good grades of smelting rock, for the mines of Clear Creek and Gilpin counties, always the largest sources of ore, were at last beginning to play out. Like J. B. Grant and Franklin Guiterman of ASARCO, Hill and Pearce tried to stimulate more energetic development by increasing returns as market conditions permitted, and on occasion they met with producers to negotiate special reduction rates. The opening of the so-called Moffat Road—the Denver, Northwestern & Pacific Railway—lowered some transportation costs, but neither lower smelting fees nor reduced freight tariffs could reverse the decline in the fortunes of the traditional minerals industry.[10]

Despite the gloomy trend, Pearce and his assistant, F. C. Knight, continued to initiate technical changes that lowered reduction costs still further. For one thing, they devised a method to calculate the precise amounts of gold, silver, and copper lost in reduction so that they could determine which slags might be profitably resmelted. For another, they developed a skimming technique to collect a very low grade of matte that in more prosperous days had been discarded. This innovation also halved the number of men required to remove slag from the reverberatory furnaces.[11]

Bad luck also accelerated Argo's decline. On the night of September 7, 1906, a blaze broke out in the refinery. Firemen brought the conflagration under control before it engulfed the entire plant, but the refinery was a total loss. Hill telegraphed word of the disaster to Boston, and Converse called an emergency meeting of the directors to consider what alternatives were available. Finally, the board decided to accept Hill's advice not to rebuild on the grounds that the firm had little prospect of making money with the old process, which was now obsolescent. Instead, Converse and his colleagues resolved to sell the furnace product to ASARCO. Over the next few years Pearce shipped the smelter's production to Omaha.[12]

The decision to sell matte to ASARCO eliminated the rationale for keeping Richard Pearce's method a corporate "secret." The scientific community had long speculated on the details of the process, and now with the refinery gone to rubble Rossiter Raymond dispatched a letter to Harold V. Pearce asking him to reveal the procedures. Both Hill and Pearce thought this a good idea, and the directors concurred. Having secured official approval, Pearce drew on his father's notes and wrote an article that Raymond had published in the *Transactions of the*

American Institute of Mining Engineers. It contained the details
Richard Pearce had been unable to discuss in his presidential
address to the society nearly twenty years before.[13]

Three years after the great fire that destroyed the refinery,
the declining fortunes of the company brought on a serious
discussion of liquidation. Talk surfaced at the annual meeting
held at Argo in April 1909. Hill and others pointed out that since
1900 five years of profitable operations had been more than
offset by four years of deficits that had left the firm with a net
loss of $60,000. In light of the unpropitious outlook for the
minerals industry, as well as the continued inability of the
smelter to get adequate supplies of ore, the directors authorized
Hill to reduce operations to the lowest possible level and dis-
charge all unnecessary workers. Hill and his associates, however, were still unwilling to make the decision to dissolve the
corporation, although events were obviously moving in that
direction.

The debate over the company's future came to a climax that
summer. In the course of his vacation Hill worried about financial problems and the question of restoring profitability. When
he returned to his Denver offices in August, he suggested to
officials in Boston that the enterprise be dissolved. Converse
agreed. He instructed Hill to convert everything possible into
cash and to purchase only those ores necessary to process what
already lay on the smelter grounds. While Hill sought legal
counsel about the procedures for liquidating a Colorado corporation, Converse called a special meeting of the directors to consider the matter formally, although there was no question about
the outcome. The board assented. On September 20, Converse
sent all stockholders a letter reviewing the financial condition of
the firm and noting its gloomy prospects. He enclosed proxies
and urged each shareholder to sign and return them so as to
empower the directors to vote for dissolution at a special meeting scheduled for early November. The response in favor of
winding up the enterprise proved overwhelming, and the directors set November 2 as the day for the stockholders to consider
the question.[14]

Shortly before the meeting, legal technicalities—ostensibly—

led to Converse's resignation as the company president. Under Colorado law the chief executive of a corporation about to be dissolved had to swear under oath that nefarious activities had not been the cause of the firm's demise. Converse was amenable to signing an affidavit about matters within his personal knowledge in Boston, but he was reluctant to sign the required document on the grounds that he was unfamiliar with daily operations at Argo. Converse must still have suspected the western management of financial deception. He volunteered to resign as president if Hill would assume the office and the concomitant responsibilities. When the resident director agreed, he was elected on November 1. And so, on the day before the stockholders met to vote on dissolution, a member of the Hill family, chief architects of the firm's destiny for more than four decades, became president of the company.[15]

All that now remained were the formalities. On November 2, 1909, Hill, Jesse D. Hale, and several associates gathered in the offices at Argo. Holding proxies on behalf of 92 percent of the stockholders, they voted to dissolve the corporation. Later that month Hill filed the official papers with the state of Colorado, and the decision became final on January 13, 1910. Operations at Argo continued for another two months until Saint Patrick's Day, March 17, when Hill's workers extinguished the fire in the last reverberatory furnace. After forty-three years in the business, the Boston and Colorado Smelting Company had processed its last ton of ore.[16]

The fires may have been out, but it took Hill and his successors another forty years to liquidate the firm's assets. At the start the trustees sold some of the houses surrounding the smelter to former employees. Firms in the metropolitan area purchased some machinery. The rest went for scrap. Wreckers demolished the furnaces, and Hill and Hale sold what slag remained to the Globe smelter and other plants for reprocessing. The Colorado & Southern Railway—the old Colorado Central line—bought several tracts of land. By 1921 all that remained of the Argo works was an empty office building with broken windowpanes, a ghostly reminder of the company's glory.

The trustees sold the remaining real estate in Denver very

slowly. Land values collapsed in the economic recession that followed World War I, and they remained low throughout the twenties and thirties and into the forties. Hill and his successors decided to hold onto the property rather than sell in such an environment. Not until after World War II, long after Hill and Converse had died, did prices rise to the point where a new generation of trustees decided to sell the land to individuals, new firms, and the Public Service Company of Colorado. In May 1950 a Denver law firm mailed a fourth and final liquidation dividend to the descendants of the original stockholders, sent a last official report to the district court, and made the unfortunate decision to have the company papers burned.[17]

Despite the problems confronting the mining industry during the late nineteenth and early twentieth centuries, another generation of entrepreneurs turned their attention toward pyritic smelting. This process had had its inception in the late 1880s and was applied early at Leadville with the formation of the old Bi-Metallic Smelting Company of Smith, Moffat, and Ballou. The company had some success, although the method failed to live up to expectations. Nonetheless, the technique had the potential of reducing profitably the vast quantities of low-grade mineral bearing silver, gold, copper, and iron. The process had particular appeal to owners of isolated mines because the smelting yielded a matte that could be shipped to refineries, whereas the ores themselves were so poor that they could not otherwise be mined, shipped, roasted, and reduced.

Except for the Bi-Metallic works in Leadville, the best-known pyritic smelter was the Carpenter plant in Golden. The firm that erected the smelter came into being in September 1900 when Joseph Berry, Ernest Le Neve Foster, and Franklin R. Carpenter organized themselves into a body corporate as the Clear Creek Mining and Reduction Company. They set the capitalization at $500,000, although whether they actually invested this amount of money cannot be determined.[18]

Berry, Foster, and Carpenter were well known in mining and smelting circles. Berry and his brother owned the Saratoga mine and other properties near Idaho Springs. Like Meyer Guggenheim, the Berrys wished to erect their own smelter to avoid the reduction charges levied by the Globe, Grant, and Argo smelters. Foster had earned a fine reputation as a mining engineer who used Denver as the hub of his activities throughout the West. Carpenter was a metallurgist long identified with the pyritic process, most notably at Deadwood, South Dakota, where he had operated the picturesquely named Deadwood and Delaware Smelting Company. He also had an excellent reputation, although that astute investigator James D. Hague thought the Deadwood facility badly managed.[19]

With the Clear Creek firm organized, the Berry brothers, Foster, and Carpenter went ahead with their plans for a new pyritic smelter. They selected Golden as the location for the works, since this would allow ore cars to run downgrade from Clear Creek and Gilpin counties, a standard reason for building in the "valley." The enterprise purchased a tract of land bordered by the Colorado & Southern Railway and let construction contracts for a plant that would have two blast furnaces with an aggregate capacity of more than 250 tons daily.

The company caused a stir in the Clear Creek ore markets, and as Carpenter pushed the smelter to completion in July 1901, its two major rivals altered their pricing schedules to meet the competition. The Boston and Colorado enterprise took action first, when it advanced its returns on certain classes of mineral. Then, as soon as Carpenter made his first purchases from local mines, ASARCO increased its pay on shipments of low-grade iron concentrates essential in the pyritic process. The struggle was on.

From the outset Carpenter and his associates confronted a host of problems. Soon after the smelter began operations, they learned that the cast-iron water jackets on the reduction furnaces were defective. Carpenter had to suspend work until the firm could install new units. When he finally resumed smelting, he found that he could not get enough ore to run at full capacity. The Berrys' Saratoga mine could not produce more than a quarter of what the plant required. To get additional supplies, Carpenter eliminated his penalty on silica, increased ore prices, and leased the Pewabic group of mines in Gilpin County, but to no avail. More mineral was not forthcoming. As a last resort, Carpenter reduced the wages of his smelter workers, but this only touched off a strike that shut down the plant. On September 11, 1903, after little more than two years in business, the Clear Creek company closed indefinitely. Rumors circulated that the plant had never earned a profit.[20]

But if Carpenter and his associates were through, the smelter was not, or at least not yet. After a year-long hiatus, it was leased to the Independent Smelting and Refining Company, a venture organized by three entrepreneurs from Denver— Marcus A. Bettman, Theodore Marx, and Robert S. Billings. From the name they chose, it appears that they hoped to profit from opposition to ASARCO, and they set the capitalization at $1,000,000, although this hardly represented real money invested in the enterprise.

Bettman and his associates tried to avoid the mistakes of their predecessors. They remodeled the plant, purchased new machinery, built more railroad track, and accumulated an ore supply, albeit a small one. But cognizant that mineral shortages had ended the career of Carpenter's group, Bettman and his colleagues acquired sampling agencies in Idaho Springs, Black Hawk, and Boulder. They also secured lower freight tariffs and then put the smelter into operation in September 1904.

From the outset operations went badly. At this stage in their development, the mines north and west of Denver just could not supply enough smelting rock to meet the needs of the Globe, Argo, and Independent plants. Bettman and his associates could never get the ores they needed, and what was more, bad roads curtailed shipments to Golden. After scarcely nine months in business, the Independent Company blew out its furnaces. Bettman tried to raise new capital in New York, but he had no success. The firm went into receivership, and the plant reverted to the Clear Creek enterprise.[21]

The smelter lay idle for another five years until 1910, when a third group of entrepreneurs decided to try their hand at pyritic reduction. The chief figure was H. A. Reidel, president of the North American Smelter and Mines Company. He and his associates had worked mining properties above Clear Creek for some time, but now they wished to integrate their operations forward into ore processing. After taking possession of the plant, they appropriated a reported $35,000 for renovation and improvements, and Reidel hired two experienced men as superintendent and manager.

Like Bettman's group, Reidel and his associates tried to ensure themselves of adequate ore supplies. They bought additional mines near Idaho Springs and Georgetown and had the good fortune to acquire the contracts of another pyritic smelter that had recently failed after a short career. Reidel's men also happened to discover $100,000 worth of ores abandoned on the grounds of the old French smelter that had worked briefly in Golden during the late 1870s. And the North American venture had fortuitously chosen to enter the reduction industry just at the time the Boston and Colorado firm closed its doors, thus removing a major competitor.

In April 1910 Reidel and his associates set their first pyritic unit in blast. They ran the works as best they could, but neither hope nor plans nor luck could overcome the diminished production of smelting ores from the mines above Clear Creek. After a little more than a year in the industry, the North American firm ceased operation late in 1911. The enterprise went into receivership when Reidel and his associates defaulted on their bonded debt, which by then was on the order of $500,000. This was the final attempt to run the Carpenter smelter. Some time later wreckers demolished the plant.[22]

To some extent the story of Clear Creek, Independent, and North American firms typified the experience of many enterprises that erected pyritic smelters. The process appealed to many mineowners, particularly those working low-grade ores in relatively isolated areas. Plants appeared at Alma, Leadville, Florence, Robinson, Ouray, Silverton, and many other camps. Yet these operations rarely, if ever, lived up to expectation. Profit margins were always narrow, and most firms failed to stimulate enough ore production to make smelting economically feasible. Except for the Bi-Metallic plant in Leadville, few pyritic plants remained in business more than several months.[23]

Despite the decline of ASARCO, the demise of the Boston and

Colorado Company, and the failure of so many pyritic outfits, there was one enterprise—a new one—that had a fair success during these troubled times. This was the Ohio and Colorado Smelting Company, which erected a plant at a key railroad junction in the mountainous country and thus reversed the long-term trend toward building major smelters on the plains.

The events that led to the construction of this plant began in 1897 with the formation of the New Monarch Mining Company. The key figures were John C. Kortz and William A. Miles, both of Cleveland, Ohio, and Timothy Goodwin, a mining man from Leadville. They set the capitalization at $1,000,000 divided into one million shares, most of which were exchanged for the outstanding stock of the Australian Mining Company, the Monarch Gold Mining Company, and other firms holding property in Leadville. Kortz took charge as president of the venture, although Goodwin directed all operations. For the next few years the New Monarch enterprise shipped its ore production either to local works or to the valley smelters, but Kortz and his associates found the arrangement unsatisfactory. They disliked the idea of sharing profits with reduction firms like ASARCO. They also had a bad experience in 1901 when the Boston Gold-Copper Company, a short-lived pyritic smelter in Leadville, went out of business owing them a reported $58,000 in returns. And so during this time Kortz and his associates explored the possibility of erecting their own reduction plant.[24]

The discussions led in 1901 to the formation of the Ohio and Colorado Smelting Company, a firm composed largely of the officers, directors, and stockholders of the mining enterprise. Kortz became president, Miles secretary, and Goodwin general manager, the same positions they held in the mining venture, thus continuing the tradition of informal integration seen throughout the reduction industry. Once they had the smelter in operation, they intended to process the entire output of their mines, but since these properties could furnish no more than about a quarter of the projected smelting capacity, Kortz and his colleagues planned to draw upon the production of mines throughout central Colorado.

Rather than build a small plant that might prove to be a

low-volume, high-cost operation—and thus doomed—Kortz and his associates drew up plans for a major smelter that could take advantage of the economies of scale. They intended to build twenty-five ore bins that would hold fifty thousand tons of smelting rock. And they would erect four blast furnaces with a capacity of six hundred tons of silver-lead mineral daily plus another two units that could process five hundred tons of copper-bearing ores. Kortz and his colleagues were about to construct a plant second in size only to the Arkansas Valley works in Leadville.

Such ambitious plans required the most careful planning, for both the location and railroad service were crucial if the firm was to have any chance of success. Kortz and his associates debated the alternatives—Denver, Pueblo, and Leadville—then decided upon Salida, a town about sixty miles south of the carbonate camp. Salida had four railroad lines that would give the plant ready access to metallurgical coal from Crested Butte and ores from the San Juan, the old Monarch district, and Leadville. Building at Salida also meant that most fuel and ore could roll downhill to the smelter, and that transportation costs would be far less than if they had to travel as far as Denver or Pueblo or across the plains to Kansas City or Omaha.

Kortz and his colleagues now bent themselves to the task of converting a paper corporation into a working industrial unit. For the plant site, they purchased an entire mesa that faced southwest toward the snowcapped peaks of Saguache and Chaffee counties. Goodwin took the responsibility for erecting the works in addition to managing the mines in Leadville. He hired construction workers, signed contracts for machinery, and attended to all details once construction began late in 1901. Work continued through the hard Salida winter. By April 1902, Goodwin's men had completed the foundations for all major buildings and were momentarily expecting the blast furnaces to arrive from the Colorado Iron Works in Denver. About this time Kortz and Goodwin decided to set July 1 as the target date for beginning the reduction of ore to bullion, but one delay after another crept into the plans, forcing the enterprise to push the time ahead to fall. Finally, Goodwin activated his smelting units, and by November he had three furnaces in blast.

The plant drew ore supplies from many districts. The New Monarch enterprise shipped its entire output to Salida, but, as expected, the mineral provided only about a fourth of what the works needed to run at capacity. Yet the firm proved itself competitive in many markets. Goodwin obtained large shipments of smelting rock from Silver Cliff, Bonanza, and Leadville. Small consignments arrived from Gunnison and Hinsdale counties. Even the old Monarch district staged a short renaissance once the smelter opened. Mines in Colorado provided the bulk of the mineral, but they could no longer provide the lead essential in the reduction process. For this Kortz and Goodwin had to look far afield. Mines in the Northwest had the required mineral, and here Kortz and Goodwin signed contracts with producers in the Coeur d'Alene and even as far away as British Columbia.[25]

Meanwhile, the failure of the Boston Gold-Copper Company presented Kortz and his associates with an opportunity to expand and integrate operations only a short time after the Salida works came on stream. The Boston firm owned a pyritic smelter that had proved none too successful, like most such plants, but Kortz and Goodwin thought that if they acquired the plant they might use it to process extremely low grade sulfides in Leadville and ship the furnace product to Salida for resmelting. With this intent Miles bought the property and assets of the Boston firm for the account of the Ohio and Colorado enterprise. Kortz, Goodwin, and Miles reorganized the venture as the Republic Smelting and Refining Company and set the works back in operation. It was reported that they intended to double the capacity from five hundred to a thousand tons daily, but if this was true they soon changed their minds. The smelter never proved renumerative. Late in 1903, after operating the plant for about a year, they decided to abandon it and confine their reduction work to Salida.[26]

Meanwhile, Kortz and Goodwin expanded the Ohio and Colorado smelter. Early in 1903 the directors appropriated a sum reported as $100,000 to erect a sulfide mill for processing the huge quantities of low-grade material shipped down the Arkansas Valley from Leadville. Goodwin pressed forward with construction during the summer and fall, and when the plant became operational it doubled the roasting capacity. Several years later the firm again increased sulfide capacity, this time through purchasing from ASARCO the rights to install the Huntington-Heberlein process, which was coming into general use throughout the industry.

Kortz and his associates also expanded their ore purchasing. Goodwin dispatched agents to mining districts in Utah and to camps in Idaho outside the Coeur d'Alene. These buyers often received an unexpected assist because some mineowners disliked the practices of ASARCO and feared the quasi-monopoly. Goodwin signed contracts with some companies so frustrated by the "smelting trust" that they wished to send their ores to Salida even though freight charges and reduction fees were somewhat higher. Other shrewd mineowners dealt with the Ohio and Colorado firm merely to keep at least one viable competitor in business. Goodwin's agents were so energetic in finding ores that he increased his work force at Salida to 250 men and reduced as much as eight hundred tons of mineral daily. The bullion he sold to the American Metal Company.

Despite its success, the company could not avoid the labor ferment of that turbulent era. In July 1903, Goodwin tried an eight-hour day for furnacemen as a concession to the hot Salida summer and as an effort to dissuade experienced employees from departing for cooler climates. In the fall he insisted upon a restoration of the twelve-hour day. This the smelter workers opposed, and two-thirds of the force went out on strike. Goodwin and his colleagues curtailed operations but continued roasting and smelting on a small scale with nonunion men who crossed picket lines. Like their rivals in ASARCO, Goodwin and his friends had the upper hand, and the strike later ended in failure with the twelve-hour day still in effect.

As Kortz and company expanded operations, they also continued the steady development of their mining properties in Leadville. In 1903 the New Monarch firm began an extensive excavation of its new Cleveland shaft while regular work went forward in the older Winnie, Lida, and New Monarch tunnels.

Over the next few years the enterprise invested large sums of money in equipment and machinery, and through this, good ore bodies, and good luck, the firm produced about one-tenth of Leadville's output.[27]

Kortz and his associates had the good fortune in 1906 to encounter a remarkably rich streak of ore in the depths of the Winnie shaft. This deposit, only ten inches in diameter, averaged 36 ounces of gold, 77 ounces of silver, and 160 pounds of copper in every ton. The vein itself extended several hundred feet, and its discovery was a lucky reward for the steady efforts to mine and process low grades of mineral. It was just this kind of chance that spurred venture capitalists, if not gamblers, to develop other properties as the mining industry faltered.[28]

In 1907, however, the prices of silver, lead, and copper declined in the wake of the financial crisis of that year, and the newer levels had a deleterious influence on the firms of Kortz and his associates. As values fell, they had to curtail operations at both the New Monarch mines and the Salida smelter, since it was no longer profitable to work the lower grades of ore. Not unreasonably, they assumed that market values would rise once the recession had run its course. As such, Kortz and his colleagues chose this time to install new equipment at their Leadville and Salida properties. The improvements aided both firms, but metals prices unexpectedly remained about the same and would not return to the 1907 level until the United States entered World War I a decade later. In this new economic environment the production of the New Monarch mines fell from two hundred to fifty tons daily, and as shipments of smelting rock fell off throughout the high country owing to prior declines and the exhaustion of ore bodies, the Salida plant found it impossible to run at full capacity.

After 1907 the Ohio and Colorado Company limped along. Kortz and his associates thought a reduction in rail tariffs between the San Juan and Salida might stimulate greater shipments from that quarter, but the hope proved false when the lines lowered their rates. The rising production of zinc-bearing mineral did help the plant, however, for Goodwin bought large quantities of residues from zinc retort smelters in Kansas and

Oklahoma. These materials helped the firm remain in business, but not even the unusually high prices for metal created by World War I could resuscitate the traditional mining industry. The market values of lead, copper, and zinc fell off when the conflict ended in 1918. When the postwar recession eliminated the shipments of zinc residues upon which the plant had come to depend, Kortz and his associates decided to dismantle the plant. The Ohio and Colorado smelter reduced its last ton of ore in 1920.[29]

The demise of the Salida smelter left ASARCO as the only important reduction enterprise in Colorado, once the heart of this industry. Since 1907 the Globe, Pueblo, Arkansas Valley, and Durango works had operated at less than full capacity owing to lower prices for silver, lead, and copper and to the continuing decline of the mining industry based on these metals. Like the Ohio and Colorado Company, the "trust" garnered additional reduction materials through shipments of residues from zinc retort smelters, and they played a large role in keeping the Globe and Pueblo works in operation.

After World War I, however, Daniel Guggenheim and his associates saw their business in Colorado contract even further, requiring new changes in ASARCO's structure. In 1919 the firm decided to halt reduction at the Globe smelter and use the plant solely for treating cadmium, thallium, and indium compounds shipped from other works. Thus Denver, the chief smelting center in the West only two decades before, ceased to be a reduction site at all. Over the next two years the firm witnessed declining shipments of zinc residues from Kansas and Oklahoma to its Pueblo smelter. This matter came to a head in June 1921 when a disastrous flood in the lower Arkansas Valley destroyed a large portion of the works. Once the waters had receded, ASARCO decided that the time had come to close this plant as well.

The demise of the industry at Denver and Pueblo left ASARCO with only the Arkansas Valley and Durango works to process the dwindling output of smelting materials from the high country. Yet now their capacity was more than enough to handle the task. During the 1920s the AV plant reworked old slag dumps in Leadville and reduced small quantities of ore mined in the central Rockies or along the eastern slope. The Durango smelter served a few mines near Telluride and elsewhere in the San Juan. Only rarely did either plant have more than one furnace in blast, so miniscule had become the shipments of smelting rock. Then came the severe economic dislocation of the Great Depression, which forced many mines in the San Juan to close down. With that ASARCO closed its Durango works in 1930.

The firm now had only the Arkansas Valley smelter to handle all operations in Colorado. The plant continued working throughout the thirties, but rarely with more than one furnace in blast. The onset of World War II failed to revive the mining industry, but the works still obtained what ores it needed to continue running on a very limited scale throughout the forties, into the fifties, and down until the sixties. Then in 1961 ASARCO decided to close the plant. Nearly a century after James E. Lyon had opened his short-lived works in the town of Black Hawk, the smelting industry in the Rocky Mountains had come to an end.[30]

Throughout its evolution the business of ore reduction followed the great themes that characterized American industrial development during the late nineteenth and early twentieth centuries. Vertical and horizontal integration, centralization, the increased use of capital, the creation of urbanized labor forces, the quest for technological advance, the rise of professional managers, the drive for lower costs, and the rationalization of all aspects of production were just as marked in the smelting industry as they were in petroleum, steel, copper, and many lines of manufacturing. Yet the smelting industry was unique in itself, and its evolution showed that the development of western mineral resources depended largely on the adaptation of European technology and the mobilization of eastern capital.

From its inception in the isolated mining camps of the Rockies, the industry had drawn on metallurgy long used in Europe. Although some men tried their own inventions or tapped inadequate methods used in the East on different types of ore, the shrewdest entrepreneurs consciously sought processes long used in the world-renowned centers of reduction. Nathaniel P. Hill was one of the first in a long line that looked abroad for proved technology. And, like Hill, many businessmen hired skilled laborers, technicians, and metallurgists to install machinery and run plants. Even after the industry had established itself, smeltermen still looked to Europe—witness the adoption of the Huntington-Heberlein process in the early twentieth century. During this time, however, European technologies continued to evolve in the United States as Americans advanced the art and science of ore reduction to process the maximum amount of mineral at the lowest possible cost.

Although the smelting centers of Europe provided much of the technology, the capital that supported the industry came primarily from financial centers east of the Rockies. Wealthy people in Boston and New York invested heavily; others in Philadelphia, Chicago, Kansas City, and St. Louis provided some money; Europeans offered smaller, less significant sums; comparatively little capital came from the Rockies or the West Coast. The stock of most firms was closely held and not listed on any exchange, although this changed somewhat late in the century, particularly after the formation of ASARCO. Throughout this time those holding the controlling interest in most smelters resided far away, as only a few major stockholders made their homes in the Rockies or even in the West. Increasingly, control became centered on New York, the financial capital of the nation.

It was this acquisition of European technology and mobilization of eastern capital to build smelters that permitted many mines to be worked and promoted the economic development of the region. In many districts the smelters stood at the crossroads of exploitation, for, given the metallurgy of the day, only they had the technology needed to recover gold, silver, copper, lead, and other metals from the different types of ore. Without the smelters some minerals might have remained untapped for decades. And in a broader perspective, the smelters stimulated agricultural development, railroad construction, coal mining, and other industries as well as the growth of cities like Denver and Pueblo.

Despite its international scope, the industry still followed an evolutionary pattern typical of "big business" in the United States during the late nineteenth century. The first enterprises, isolated as they were, intended to serve a small local market, although they had to sell their bullion, matte, and metal in a world market. Entry into the industry was relatively easy, the capital requirements small, and hard competition normal. Later, as individual companies grew larger through mergers and internal expansion, they invaded more distant territories and converted themselves into interregional and international processors. By this time, however, entry had grown far more difficult and the capital requirements larger. The major firms then grew more energetic in their efforts to reduce competition through pools and even larger mergers, the last of which created a holding company that took control of almost the entire business.

Technologically, the industry evolved according to science and engineering rather than the rule-of-thumb techniques of former times. Arthur S. Dwight once characterized the early days as the age of "muscular metallurgy," but this era disappeared as metallurgists adapted more systematic techniques, defined desirable slag types, and standardized procedures. Americans also put their characteristic stamp of large-scale, low-cost operations on the older methods of Europe.

Ultimately, of course, the industry depended on the mines. Many smelters prospered when they had high-grade minerals to reduce, but as values declined, ores grew more complex, and the price of metals fell, the smelters suffered with the mines, since the two industries lived together in symbiosis. Finally, when the grades of ore fell below the costs of smelting, miners had to turn to other technologies or else close down—and they did both. In either case, the great plants in Denver, Pueblo, Leadville, and Durango had to close their doors and go the way of the wrecker. Even though these works disappeared, the American Smelting and Refining Company remained because the Guggenheims and their associates maintained plants in other cities like Omaha and El Paso and converted the firm into a processor of many metals, a transition already under way by the time of the great merger in 1899. By the onset of World War I, the annual value of the gold and copper marketed by ASARCO exceeded the value of silver and lead, on which the predecessor companies had evolved. And as the years passed, the enterprise became primarily identified with the copper industry and remains so to this day.

Not much remains in the Rockies to remind a passerby of the once-great industry there. No picturesque ruins dot the high country, no bronze statues memorialize the smeltermen, no legends recount past glories. A few slag dumps lie here, an abandoned furnace there, an isolated smokestack pierces the skies somewhere else, but nothing more. A few place names remain—Argo and Globeville in Denver, Eilers Street in Pueblo, Harrison Avenue in Leadville—but not many can recount the origin of the names. The legacy of the business lies elsewhere, primarily in the American Smelting and Refining Company, now officially known as ASARCO, which operates in many locales but only marginally in the Rocky Mountains, once the heart of the industry that created the firm.

Notes

Full details of publication for sources cited are given in the Bibliography.

Abbreviations

CHS	Colorado Historical Society
CM	Commonwealth of Massachusetts
CSA	Colorado State Archives
CU	University of Colorado
DPL	Denver Public Library
EMJ	*Engineering and Mining Journal*
FRCD	Federal Records Center, Denver
HEH	Henry E. Huntington Library and Art Gallery
HU	Harvard University
MSP	*Mining and Scientific Press*
NA	National Archives and Records Service
NYPL	New York Public Library
YU	Yale University

Chapter 1

1. Charles W. Henderson, *Mining in Colorado*, pp. 1–17.
2. Rodman Wilson Paul, *Mining Frontiers of the Far West*, 1848–1880, pp. 111–14.
3. Ovando J. Hollister, *Mines of Colorado*, pp. 60–63; Henderson, *Mining in Colorado*, pp. 1–7.
4. Hollister, *Mines of Colorado*, pp. 227–39; Henderson, *Mining in Colorado*, pp. 1–6, 36–37.
5. Henderson, *Mining in Colorado*, passim; Paul, *Mining Frontiers*, pp. 115–16.
6. Edson S. Bastin and James M. Hill, *Economic Geology of Gilpin County and Adjacent Parts of Clear Creek and Boulder Counties, Colorado*, p. 105.
7. Henderson, *Mining in Colorado*, p. 29; Paul, *Mining Frontiers*, pp. 3–7, 19–20, 31–32.
8. Hollister, *Mines of Colorado*, pp. 331–39.
9. Ibid., pp. 103, 110–12; Bastin and Hill, *Economic Geology of Gilpin County*, p. 134.
10. Hollister, *Mines of Colorado*, pp. 64–65.
11. Ibid., pp. 136–39; Joseph E. King, *A Mine to Make a Mine*, pp. 3–21.
12. Rossiter W. Raymond, "Report upon Various Desulphurizers, in Their Adaption to the Sulphurets of Colorado," Raymond and Adelberg Collection; Rossiter W. Raymond, *Statistics of Mines and Mining in the States and Territories West of the Rocky Mountains* (1870), pp. 348, 357; there are eight reports in the Raymond series from 1868 to 1875, each of which is cited by annual report for a given year.
13. Nathaniel S. Keith to Anna Keith, Black Hawk, January 25, 1863, Keith Collection; Raymond, *Annual Report, 1870*, pp. 356–59.
14. A. W. Hoyt, "Over the Plains to Colorado," pp. 1–21.
15. Testimony of Nathaniel P. Hill in U. S. House of Representatives, *Report of the Industrial Commission*, House Doc. 181, 57 Cong., 1st sess., serial 4342, vol. 12, 1901, p. 380.
16. Hill to Alice Hill, Denver, June 19, 1864; unless otherwise indicated, all letters cited in this chapter are from the N. P. Hill Collection, CHS. Thomas L. Karnes, *William Gilpin, Western Nationalist*, pp. 302–5.
17. Charles M. Perry, "Chemistry and Its Teachers in Brown University," Brown University Archives, p. 12; *Catalogue of the Officers*

and Students of Brown University 1854–1855 [*through*] *1864–1865*, passim; B. W. Steele, "Hon. Nathaniel P. Hill," pp. 431–33.

18. Perry, "Chemistry and Its Teachers in Brown University," Brown University Archives, p. 13; newspaper clippings in the N. P. Hill Collection, CHS; *Examination by Chemical Analysis and Otherwise, of Substances Emptied into the Public Waters of the State, from Gas and Other Manufactories, Sewerage and Other Sources . . . :* and *Transactions of the Rhode Island Society for the Encouragement of Domestic Industry in the Year 1859*, pp. 43–47.

19. Hill to the executive board of Brown University, n.d., Brown University Archives. Hill wrote the letter between January and April 1864; he did not elaborate on the two offers he rejected; and Hill to Alice Hill, Denver, June 30, 1864.

20. Hill to Uriah W. Lawton, Brown University, March 9, 1857, Brown University Archives. Hill to Belle Hill, Denver, June 15, 1864; to Alice Hill, Denver, June 19, July 31, and September 4, 1864; and Costilla, New Mexico, July 21, 1864.

21. Hill to Alice Hill, Denver, June 14, 1864; to Belle Hill, June 15, 1864; and unidentified newspaper clipping, N. P. Hill Collection, CHS.

22. Hill to Alice Hill, Denver, June 14, 1864.

23. Ibid., Central City, June 23, 1864; Karnes, *William Gilpin*, pp. 302–5.

24. Hill to Belle Hill, Denver, June 15, 1864; Hill to Alice Hill, Denver, June 14, 19, 23, 30, 1864. Brastow's earlier connection with Reynolds can be seen in Henry B. Brastow to William H. Reynolds, Havana, April 1, 1863, and several other communications listed under William H. Reynolds, Twelve Facsimile Letters, Rhode Island Historical Society.

25. Hill to Alice Hill, Denver, June 30; fifty-five miles south of Denver, July 5; Sangre de Cristo Pass, July 10, 1864.

26. Ibid., Denver, July 30, 1864; Hill to Belle Hill, Denver, August 1, 1864.

27. Hill to Alice Hill, Denver, August 8 and September 4; Pueblo, September 16, 1864.

28. Ibid., Central City, October 3, 1864.

29. Ibid., Central City, October 16, 1864; newspaper clipping in the N. P. Hill Collection, CHS.

30. Aborn to Gilpin, Providence, January 11, 1865, in William Blackmore, *Colorado: Its Resources, Parks and Prospects as a New Field for Emigration . . .* , p. 204; Karnes, *William Gilpin*, pp. 308–9.

31. Hill to Alice Hill, Central City, October 16, 1864; *Brown Alumni Monthly* (December 1900), p. 77.

32. The incorporators of the Hill Gold Mining Company were James T. and Frank A. Rhodes, Edward Pearce, Alexis C. Caswell, Jabez C. Knight, Moses B. Lockwood, Frank Mauran, and F. H. Richmond, copy of certificate of incorporation, CSA.

33. Hill to Alice Hill, Cleveland, February 19; Saint Joseph, February 22; Atchison, February 24; Alkali Lake, March 5; and Central City, March 10, 1865.

34. Ibid., Central City, March 10, 19, 22, 1865.

35. Ibid., London, March 29, 1866.

36. Lyon to Matt [?], New York, May 17, 1865, and other documents in box 3, Teller Collection, CU.

37. The *Daily Miner's Register* is quoted in Joel Parker Whitney, *Silver Mining Regions of Colorado, with Some Account of the Different Processes Now Being Introduced for Working the Gold Ores of That Territory*, p. 19; for Lyon's purchases, see scattered entries in his Record Book and Journal, CHS.

38. *Rocky Mountain News*, January 2, 10, April 2, 1866.

39. Hill to the executive board, 1862, Brown University Archives; and Hill to Alice Hill, London, March 29, 1866.

40. Hill to Alice Hill, Liverpool, March 14; London, March 16, 19, 1866; and Hill to George Jarvis Brush, Providence, February 15, 1866, box 3, Brush Collection.

41. Hill to Alice Hill, Swansea, April 3, 6; London, April 14, 1866; Roberts, "Development and Decline of the Non-ferrous Metal Smelting Industries in South Wales," in W. E. Minchinton, ed., *Industrial South Wales, 1750–1914: Essays in Welsh Economic History*, pp. 121–60.

42. Ledger Accounts, New York & Colorado Smelting Works, box 3, Teller Collection, CU; Lyon's Record Book, June 9, 1866, and passim, CHS; *Rocky Mountain News*, April 2, 1866; *Pioneer Smelting Company of Colorado*; Herrmann to Lyon, Central City, July 8, 1866, in the *Rocky Mountain News*, August 13, 1866; Lyon's Journal, March, 1867, CHS; Hollister, *Mines of Colorado*, pp. 355–56; Raymond, *Annual Report, 1870*, p. 359.

43. *Rocky Mountain News*, June 20, 1866; Herrmann to Lyon, Central City, July 8, 1866, in the *Rocky Mountain News*, August 13, 1866; Robert O. Old to Henry C. Justice, London, November 9, 1866, in the *Rocky Mountain News*, December 8, 1866; Hill to the editor, Black Hawk, July 27, 1869, in the London *Mining Journal*, 39 (September 1869: 667; Hollister, *Mines of Colorado*, pp. 158–60; Bayard Taylor, *Colorado: A Summer Trip*, p. 67; Samuel Cushman and J. P. Water-

man, *Gold Mines of Gilpin County, Historical, Descriptive, and Statistical*, p. 103.

44. Certificate of incorporation, book 56, p. 260, CM.

45. Converse Papers; certificates of condition, Boston Rubber Shoe Manufacturing Company, CM; *National Cyclopaedia of American Biography* 12:166; Ernest Cummings Marriner, *History of Colby College*, passim; Hill to Alice Hill, Atchison, February 24, 1865; *Historical Statistics of Brown University, 1764–1934*, pp. 9–10; Records of the Trustees, Brown University Archives; *Catalogue of the Officers and Students of Brown University*, passim; J. Ross Browne and James W. Taylor, *Reports upon the Mineral Resources of the United States*, p. 45.

Chapter 2

1. Samuel Cushman and J. P. Waterman, *Gold Mines of Gilpin County, Historical, Descriptive, and Statistical*, p. 103; James D. Hague, *Mining Industry*, pp. 577–86; R. Pearce, "Progress of Metallurgical Science in the West," p. 56; *Colorado Transcript* (Golden), June 26, September 25, 1867.

2. Testimony of Beeger in *Keyes and Arents vs. Grant and Grant*, U.S. Circuit Court, docket 613, FRCD; testimony of N. P. Hill in U.S., House, *Report of the Industrial Commission*, vol. 12, p. 377.

3. Notebook no. 5, pp. 26–32, Hague Papers in Natural Resources Manuscript Group, YU; Cushman and Waterman, *Gold Mines of Gilpin County*, p. 103; *Colorado Transcript* (Golden), June 26, September 25, October 23, 1867.

4. Hague, *Mining Industry*, p. 580; Egleston, "Boston and Colorado Smelting Works," p. 281; Edward D. Peters, Jr., *Principles of Copper Smelting*, pp. 23 ff.; *MSP* 19 (July 1969): p. 22.

5. Hague, *Mining Industry*, pp. 581–82; Peters, *Principles of Copper Smelting*, pp. 168–69.

6. Hague, *Mining Industry*, pp. 582–86.

7. Ibid., pp. 579–83; Hill to the editor of the *Central City Register*, n.p., n.d., printed in Raymond, *Annual Report, 1871*, pp. 345–47; notebook no. 5, p. 40, Hague Papers, YU; and *MSP* 22 (January 1871):8.

8. Hague, *Mining Industry*, pp. 562–66, 584–85; Raymond, *Annual Report, 1870*, p. 372.

9. Hague, *Mining Industry*, pp. 496–97, 562–66, 584–85.

10. *Daily Miner's Register* (Black Hawk), June 18, 19, 1868.

11. Raymond, *Annual Report, 1870*, p. 354; letter fragment written in 1871 by Anton Eilers, Adelberg and Raymond Collection.

12. William Larned to George W. Heaton, Central City, November

27, 1871, box 4, Teller Papers, CU; Eilers's letter fragment, Adelberg and Raymond Collection; Hague, *Mining Industry*, p. 580; Raymond, *Annual Report, 1872*, p. 289; *MSP* 23 (August 1871):83; 23 (September 1871):133.

13. Raymond, *Annual Report, 1872*, pp. 265, 270–71.

14. Robert O. Old to George W. Heaton, Georgetown, September 7, 1870; Vivian & Sons to Heaton, Swansea, September 21, 1870; William R. Kennedy to Heaton, Central City, October 26, 1870; Thomas Jennings to Heaton, Bald Mountain, Colorado, January 29, 1871, box 4, Teller Papers, CU; Raymond, *Annual Report, 1872*, pp. 265, 270–71; *Annual Report, 1873*, p. 287; *Annual Report, 1874*, p. 388; Clark Christian Spence, "Robert Orchard Old and the British and Colorado Mining Bureau," p. 120; Rickard, "Richard Pearce,"pp. 404–6.

15. West to George W. Heaton, Black Hawk, September 18, 1870, and March 8, 1871, box 4, Teller Papers, CU; Charles W. Henderson, *Mining in Colorado*, p. 38, 105–6; *MSP* 33 (November 1871):309.

16. Hill to the editor of the *Central City Register*, n.p., n.d., in Raymond, *Annual Report, 1871*, pp. 345–47; Hill to the editor, Black Hawk, July 27, 1869, in the London *Mining Journal* 39 (September 1869):667; Spence, "Robert Orchard Old," pp. 51–52; and John H. Tice, *Over the Plains and on the Mountains; or, Kansas and Colorado, Agriculturally, Minerologically, and Aesthetically Described*, pp. 111–13.

17. *Rocky Mountain News*, November 13, 1869; Raymond, *Annual Report, 1870*, p. 372.

18. Henry R. Wolcott to father, Black Hawk, June 19, 1870, in Thomas Fulton Dawson, *Life and Character of Edward O. Wolcott Late a Senator of the United States From the State of Colorado*, pp. 81–82; *MSP* 20 (April 1870):212.

19. Raymond, *Annual Report, 1870*, p. 332; *Annual Report, 1871*, pp. 365–66.

20. For information on the association, see the James V. Dexter Collection.

21. Raymond, *Annual Report, 1872*, p. 296; Dawson, *Edward O. Wolcott*, p. 80; Hale, "First Successful Smelter in Colorado," p. 163; *MSP* 25 (November 1872):117.

22. Henderson, *Mining in Colorado*, pp. 83–89; "Value of the Gold, Silver, and Copper . . . ," box 38, C. Hill Collection, CHS; Edward D. Peters, Jr., "The Mount Lincoln Smelting Works," pp. 310–14; see also the Dexter Collection.

23. Dawson, *Edward O. Wolcott*, p. 80; Jesse D. Hale, "The First Successful Smelter in Colorado," p. 163; Cushman and Waterman,

Gold Mines of Gilpin County, p. 103; Rickard, "Richard Pearce," p. 408; Rossiter W. Raymond, "Henry Williams," p. 8.

24. Robert G. Athearn, *Union Pacific Country*, pp. 132–33; Robert G. Athearn, *Rebel of the Rockies: A History of the Denver and Rio Grande Western Railroad*, pp. 15, 25, 43; M. C. Poor, *Denver South Park & Pacific, A History of the Denver South Park & Pacific Railroad and Allied Narrow Gauge Lines of the Colorado & Southern Railway*, passim; Traxler, "Colorado Central Railroad," pp. 44–47, 51–53; O. L. Baskin & Company, *History of the City of Denver, Arapahoe County, and Colorado*, p. 248.

25. Hague, *Mining Industry*, p. 577; Rickard, "Richard Pearce," p. 408; certificate of condition, Boston and Colorado Smelting Company, 1871, CM

26. Certificates of condition, 1873 and 1874, CM

27. Rickard, "Richard Pearce," p. 407.

28. Percy to George Jarvis Brush, London, April 7, 1875, Brush Collection, YU; and Rickard, "Richard Pearce," pp. 404–7.

29. Thomas Egleston, "Boston and Colorado Smelting Works," pp. 285–98; *MSP* 45 (October 1882):266; 48 (March 1884):195.

30. Rickard, "Richard Pearce," p. 407; Egleston, "Boston and Colorado Smelting Works," pp. 295–98; H. V. Pearce, "The Pearce Gold-Separation Process," p. 723.

31. Rickard, "Richard Pearce," p. 407.

32. Pearce, "Pearce Gold-Separation Process," pp. 723–24; Hale, "First Successful Smelter in Colorado," p. 163.

33. Pearce, "Pearce Gold-Separation Process," pp. 726–27; Rickard, "Richard Pearce," p. 408.

34. Rickard, "Richard Pearce," p. 408; Hiram W. Hixon, *Notes on Lead and Copper Smelting and Converting*, pp. 18–19.

35. Persifor Fraser, "Colorado Notes," p. 140; notebook no. 13 (1884), pp. 121–22, 138–39, Weitbrec Collection, CHS; H. O. Hofman, *Metallurgy of Copper*, p. 298; *MSP* 45 (November 1882):310.

36. Pearce, "Pearce Gold-Separation Process," p. 274; Hale, "First Successful Smelter in Colorado," p. 165; H. Van F. Furman, "The Metallurgy of Copper," p. 90.

37. Egleston, "Boston and Colorado Smelting Works," p. 283; Rickard, "Richard Pearce," p. 407.

38. Hague, *Mining Industry*, passim; Rossiter W. Raymond, "Remarks on the Precipitation of Gold in a Reverberatory Hearth," pp. 320–22; Egleston, "Boston and Colorado Smelting Works," pp. 276–98.

39. Egleston, "Boston and Colorado Smelting Works," plate 7;

Cushman and Waterman, *Gold Mines of Gilpin County*, p. 103; Raymond, *Annual Report, 1875*, pp. 294–95; *EMJ* 23 (May 1877):319; 24 (September 1877):189; 24 (December 1877):458.

40. R. B. Potter Interview, May 24, 1886, Bancroft Collection, University of California at Berkeley, copy in Norlin Library, University of Colorado; *Denver Daily Times*, May 5, 1878.

41. Henderson, *Mining in Colorado*, pp. 88–89; Frank Fossett, *Colorado. Its Gold and Silver Mines, Farms and Stock Ranges, and Health and Pleasure Resorts*, p. 243; Raymond, *Annual Report, 1872*, pp. 265–66, 288–89; *Annual Report, 1873*, pp. 284–87; *Annual Report, 1874*, p. 361.

42. Sidney Glazer, ed., "A Michigan Correspondent in Colorado, 1878," p. 215. The journalist, an employee of the *Michigan Christian Herald*, identified himself only as "T." *EMJ* 29 (September 1877):237.

43. Rickard, "Richard Pearce," p. 408; Fossett, *Colorado*, p. 243; K. Ross Toole, "A Study of the Anaconda Copper Mining Company," p. 12.

44. Raymond, *Annual Report, 1875*, p. 295; Cushman and Waterman, *Gold Mines of Gilpin County*, p. 104.

45. Hill to Teller, Black Hawk, November 14, 1877, Teller Collection, DPL.

46. *United States vs. Nathaniel P. Hill and the Boston and Colorado Smelting Company*, U.S. Circuit Court, docket 121, FRCD; Hill to Teller, Black Hawk, January 2, 1878, Teller Collection, DPL

47. Rickard, "Richard Pearce," p. 408; Ralph Newton Traxler, Jr., "Some Phases of the History of the Colorado Central Railroad, 1865–1885," pp. 70–71; *EMJ* 25 (March 1878):162–63; Athearn, *Rebel of the Rockies*, p. 25.

48. Henderson, *Mining in Colorado*, pp. 69, 88–89; "Value of the Gold, Silver, and Copper . . . ," box 37, C. Hill Papers.

Chapter 3

1. Ovando J. Hollister, *Mines of Colorado*, pp. 252–53; Charles W. Henderson, *Mining in Colorado*, p. 32.

2. Notebook no. 4, pp. 143–45, 157, Hague Papers, YU; James D. Hague, *Mining Industry*, pp. 617–18; Henderson, *Mining in Colorado*, p. 33.

3. Notebook no. 4, pp. 154, 157, notebook no. 6, p. 28, Hague Papers, YU; Hague, *Mining Industry*, p. 620.

4. Collom to Philo T. Shelton in notebook no. 5, p. 137, Hague Papers, YU; *MSP* 19 (October 1869):230; 19 (December 1869):374.

5. Henderson, *Mining in Colorado*, p. 32.

6. Hollister, *Mines of Colorado*, p. 257; Hague, *Mining Industry*, p. 609.

7. Raymond, *Annual Report, 1870*, p. 374; Hollister, *Mines of Colorado*, pp. 256–57.

8. Hague, *Mining Industry*, pp. 610–11; Raymond, *Annual Report, 1870*, pp. 375–76; H. O. Hofman, *Metallurgy of Lead and the Desilverization of Base Bullion*, pp. 198–203.

9. Notebook no. 4, pp. 126–29, 139, Hague Papers, YU; ore schedule of the Georgetown Silver Smelting Company, April 27, 1868, box 27, Hague Papers, HEH.

10. Notebook no. 4, pp. 129–31, Hague Papers, YU; Hague Papers, HEH, box 17; Hague, *Mining Industry*, pp. 607–8.

11. *First Annual Report of the President and Directors of the Brown Silver Mining Company, of Colorado, for the Year Ending March 2, 1868*; notebook no. 4, pp. 88–89, Hague Papers, YU; and Richard Pearce, "Progress of Metallurgical Science in the West," p. 57.

12. Watson to George B. Walker, Salt Lake City, December 21, 1871, January 5, 19, 1872, box 7, Teller Papers, CU; Raymond, *Annual Report, 1870*, pp. 374–75; Pearce, "Progress of Metallurgical Science in the West," pp. 56–57; *Colorado Miner*, October 14, 1869, p. 1; *MSP* 18 (May 1869):310.

13. Clark C. Spence, *Mining Engineers and the American West: The Lace-Boot Brigade, 1849–1933*, pp. 25–38, 240; Arthur S. Dwight, "Reminiscences," in Thomas A. Rickard, ed., *Rossiter Worthington Raymond: A Memorial*, pp. 64–67.

14. Otto H. Hahn, Anton Eilers, and Rossiter W. Raymond, "The Smelting of Argentiferous Lead Ores in Nevada, Utah, and Montana," pp. 91–131.

15. Hofman, *Metallurgy of Lead*, pp. 298–307; Edward D. Peters, Jr., "The Mount Lincoln Smelting Works," p. 311; Eilers, "Coke from Lignites," pp. 101–2.

16. Hofman, *Metallurgy of Lead*, pp. 305–19; Walter Renton Ingalls, *Lead and Zinc in the United States, Comprising an Economic History of the Mining and Smelting of the Metals and the Conditions Which Have Affected the Development of the Industries*, pp. 37–46.

17. J. F. Cannon to Frederick Roeser, Denver, March 13, 1917; R. P. Reynolds to L. G. Eakins, Denver, December 13, 1913; "The Manufacture of Metallic Cadmium at the Globe Plant," report prepared for K. S. Guiterman, October 9, 1911, Frederick Roeser Collection; Hofman, *Metallurgy of Lead*, passim; Ingalls, *Lead and Zinc*, passim.

18. Hofman, *Metallurgy of Lead*, pp. 305–19; Ingalls, *Lead and Zinc*, pp. 37–46.

19. Hahn, Eilers, and Raymond, "Smelting of Argentiferous Lead Ores," pp. 91–131; Rodman Wilson Paul, *Mining Frontiers of the Far West, 1848–1880*, pp. 102–5, 152–55; Ingalls, *Lead and Zinc*, p. 32.

20. Collom to Philo T. Shelton in notebook no. 5, p. 137, Hague Papers, YU; Raymond, *Annual Report, 1873*, p. 357; Peters, "Mount Lincoln Smelting Works," pp. 310–14.

21. Raymond, *Annual Report, 1873*, pp. 396–99; Colorado, 2:389 and 3:6, 156, Dun Records.

22. Raymond, *Annual Report, 1873*, pp. 305–6; Eleanor Bradley Peters, *Edward Dyer Peters* (1849–1917), pp. 60–66.

23. Peters, "Mount Lincoln Smelting Works," pp. 310–14; Raymond, *Annual'Report, 1873*, p. 398; Colorado, 2:389, Dun Records; *MSP* 25 (November 1872):325; 25 (December 1872):370.

24. E. B. Peters, *Edward Dyer Peters*, passim.

25. Joseph L. Jernegan, "The Whale Lode of Park County, Colorado Territory," pp. 352–56; Joseph L. Jernegan, "Notes on a Metallurgical Campaign at Hall Valley, Colorado," pp. 560–75; Clark C. Spence, *British Investments and the American Mining Frontier:1860–1901*, p. 249. The quotation is in Raymond, *Annual Report, 1875*, p. 358.

26. Jernegan, "Whale Lode," pp. 352–56; Jernegan, "Notes on a Metallurgical Campaign," pp. 560–75. The quotations are in the latter article, p. 575.

27. Charles Holland to Harper M. Orahood, Chicago, November 27, 1875, January 3, 1876, box 3, Orahood Papers; Raymond, *Annual Report, 1871*, p. 363; *Annual Report, 1872*, pp. 285–86; *Annual Report, 1875*, pp. 315–17; Colorado, 2:339 and 3:105, Dun Records.

28. Ingalls, *Lead and Zinc*, passim; Raymond, *Annual Report, 1871*, passim; *MSP* 28 (March 1869):182.

29. Records of the Omaha Smelting and Refining Company, CSA; Nash to Hague, Omaha, May 12, 1879, box 15, Hague Papers, HEH; Lowe to Bristol & Daggett, Omaha, September 5, 1871, Bristol Family Collection; Nebraska, 3:226, and New Jersey, 21:102–3, Dun Records; *MSP* 21 (July 1870):80.

30. Joseph C. Mather to Bristol & Daggett, Salt Lake City, August 5, 1872; E. P. Vining to Ellsworth Daggett, Omaha, July 29, 1872; Lowe to Bristol & Daggett, Omaha, September 5, 1871; Eugene Stuart Bristol to William B. Bristol, Bingham Canyon, Utah, October 26, 1871, Bristol Family Collection; *MSP* 22 (July 1871):405.

31. Lowe to Lewis H. Bristol, Omaha, October 17, 1871; Rustin to E.

S. Bristol, Omaha, December 12, 1871, May 2, 11, 1872; Gustav Billing to Bristol & Daggett, Salt Lake City, December 2, 1872; certificates of assay, May 4, 10, 1872, Bristol Family Collection, YU; *MSP* 39 (September 1879):149; 40 (May 1880):341.

32. Nash to Hague, Omaha, May 12, 1879, and "The 'Atlanta' Mine," by W. H. Pettet, September 7, 1892, Hague Papers, HEH; *Francis E. Everett and the Omaha Smelting and Refining Company vs. the Hukill Gold and Silver Mining Company*, U.S. Circuit Court, docket 213, FRCD; Nebraska, 3:226, Dun Records.

33. Statement of Edwin Harrison, November 13, 1899; Edwin Harrison to James Harrison, Santa Fe, March 5, 1862; Louis Agassiz to whom it may concern, Cambridge, April 4, 1858; Louis Agassiz to Edwin Harrison, January 24, 1858, May 28, June 19, 1873; and newspaper clippings, Harrison Family Collection.

34. Edwin Harrison to Joseph Bogy, St. Louis, July 18, 25, 1872, Rosemary Bogy Collection; statement of Edwin Harrison in Bryan Obear Scrapbooks; "The Disappearance of Iron Mountain: Story of the Rise and Fall of a Great Industry in Missouri," in the *Globe-Democrat* (St. Louis), October 16, 1898.

35. Records of the St. Louis Smelting and Refining Company, CSA; Missouri, 43:148, Dun Records; *Industries of St. Louis, Her Relation as a Center of Trade. Manufacturing Establishments, and Business Houses*, pp. 101–2.

36. Augustus Steitz to George W. Maynard, Denver, March 10, 1866, in *Prospectus of the Belle Monte Furnace Iron and Coal Company of Boulder County, Colorado*; Dwight, "Reminiscences," in Rickard, ed. *Rossiter Worthington Raymond*, pp. 64–67; *Republican* (St. Louis), January 1, 1874; Raymond, *Annual Report, 1874*, p. 457; Fahey, *D. C. Corbin and Spokane*, pp. 10–13; *MSP* 18 (June 1869):409.

37. H. S. Sanders to Bristol & Daggett, n.p., May 25, 1872, Bristol Family Collection; Missouri, 43:148, Dun Records; *Republican* (St. Louis), January 1, 1874.

38. Missouri, 43:148, Dun Records; The quotation is in the *Republican* (St. Louis), January 1, 1874.

39. Henderson, *Mining in Colorado*, passim.

Chapter 4

1. Ovando J. Hollister, *Mines of Colorado*, pp. 311–17; Don Griswold and Jean Griswold, *Carbonate Camp Called Leadville*, pp. 2–4.

2. Duane A. Smith, *Horace Tabor: His Life and the Legend*, pp. 19–23, 26; Griswold and Griswold, *Carbonate Camp*, pp. 2–4.

3. Hollister, *Mines of Colorado*, pp. 316–20; Samuel F. Emmons, *Geology and Mining Industry of Leadville, Colorado*, p. 10.

4. Emmons, *Leadville*, p. 10; Smith, *Horace Tabor*, p. 26.

5. Emmons, *Leadville*, p. 10; N. P. Hill to Alice Hill, Central City, October 3, 1864, N. P. Hill Collection, CHS.

6. Emmons, *Leadville*, p. 10; Charles W. Henderson, *Mining in Colorado*, pp. 133–35; Western Territories, 1:14, and Colorado, 2:360, Dun Records.

7. Henderson, *Mining in Colorado*, pp. 133–35; Smith, *Horace Tabor*, pp. 37–38.

8. Henderson, *Mining in Colorado*, pp. 133–35; Raymond, *Annual Report, 1876*, p. 314; *Rocky Mountain News*, November 29, 1873.

9. Records of the Malta Smelting and Mining Company, CSA; Emmons, *Leadville*, p. 11; Raymond, *Annual Report, 1876*, p. 314; *EMJ* 22 (September 1876):172; 22 (December 1876):423.

10. *EMJ* 22 (September 1876):172; 22 (December 1876):423.

11. Hague to C. W. Colehour, Leadville, August 5, 1879, box 20, Hague Papers, HEH; Emmons, *Leadville*, p. 12; Griswold and Griswold, *Carbonate Camp*, pp. 22–24.

12. Griswold and Griswold, *Carbonate Camp*, passim.

13. Emmons, *Leadville*, p. 12.

14. A. Middlebrook to Harrison, Denver, June 21, 1877, Harrison Papers; C. W. Dana to the editor, Oro City, n.d., in *EMJ* 23 (January 1877):55; 23 (May 1877):352; Maurice Hayes to the editor, Oro City, March 19, 1877, in *EMJ* 23 (March 1877):205; Missouri, 43:148, Dun Records; *A. R. Meyer & Company Ore Milling and Sampling Company vs. J. B. Dixon*, U.S. Circuit Court, docket 374, FRCD.

15. A. Middlebrook to Harrison, Denver, June 21, 1877, Harrison Papers; L. A. Kent, *Leadville. The City. Mines and Bullion Product. Personal Histories of Prominent Citizens. Facts and Figures, Never Before Given to the Public*, pp. 94–95; *Saint Louis Smelting and Refining Company vs. H. P. Parlon and D. Shaw*, U.S. Circuit Court, docket 292, FRCD. For other lawsuits by the enterprise, see dockets 291, 309–28, and scattered cases from 332 to 366; *MSP* 20 (May 1870):340.

16. *EMJ* 25 (April 1878):240, 293; Kent, *Leadville*, pp. 94–95.

17. Emmons, *Leadville*, p. 14; Kent, *Leadville*, pp. 22–23; U.S., Congress, House, Committee on Ways and Means, *Revision of the Tariff: Hearings before the Committee on Ways and Means, 1889–90*, H. Misc. Doc. 176, serial set 2774, 51st Cong., 1st sess., 1889, p. 1277.

18. Emmons, *Leadville*, p. 13; Smith, *Horace Tabor*, pp. 71–76.

19. Emmons, *Leadville*, pp. 13–14; Smith, *Horace Tabor*, pp. 76–77.

20. *EMJ* 23 (March 1877):170; 25 (March 1878):221; *MSP* 38 (April 1879):213, 230.

21. Emmons, *Leadville*, pp. 14–15; *MSP* 39 (November 1879):284.

22. *EMJ* 25 (March 1878):221, 223; 25 (April 1878):293; 25 (June 1878):445; Kent, *Leadville*, passim; Joseph E. King, *A Mine to Make a Mine*, pp. 82–115.

23. Kent, *Leadville*, passim; Henderson, *Mining in Colorado*, p. 89.

24. Henderson, *Mining in Colorado*, p. 89; Kent, *Leadville*, p. 115; Emmons, *Leadville*, p. 626.

25. *EMJ* 25 (April 1878):293.

26. *Meyer et al. vs. Dixon*, docket 374, FRCD.

27. Interview with Franz Fohr, Bancroft Collection; Wood, "I Remember," Wood Papers.

28. *EMJ* 26 (August 1878):154; Kent, *Leadville*, pp. 105–6, 112; Walter Renton Ingalls, "Franz Fohr," *EMJ* 108 (November 1919):828–29; *Meyer et al. vs. Dixon*, docket 374, FRCD.

29. Hynes to C. B. Beitzel, Leadville, January 1, 1879; Hynes to Harrison, May 28, 1879; Hynes to Edwin Harrison & Company, July 11, 16, 1879, Harrison Papers; Kent, *Leadville*, pp. 94–95; *EMJ* 26 (September 1878):208.

30. John Evans to Harrison, Chicago, February 7, 1878; Tabor to Harrison, Oro City, February 15, 1878, and Leadville, March 11, 1878; Henry B. Yelita to Harrison, Alpine, Colorado, October 14, 1878; legal document signed by Harrison and William Teague, August 30, 1879, Harrison Papers; *Argentine Mining Company of St. Louis of the State of Missouri vs. the Adelaide Consolidated Mining and Smelting Company*, U.S. Circuit Court, docket 214, FRCD; *EMJ* 26 (September 1878):208.

31. *EMJ* 98 (November 1911):1004.

32. The quotation is in the "Report of James B. Grant, Leadville, March 16, 1878," in *Pueblo & Oro Railroad*, p. 17. See also James Grant to Henry Amy & Company, Davenport, Iowa, March 23, 1878; ibid., p. 23.

33. Iowa, 1:185, Dun Records; James Grant to Henry Amy & Company, Davenport, Iowa, March 23, 1878, in *Pueblo & Oro Railroad*, p. 23.

34. *EMJ* 25 (June 1878):405; Kent, *Leadville*, pp. 101–4; *James B. Grant vs. Council Bluffs Iron Works*, U.S. Circuit Court, docket 417, FRCD.

35. *EMJ* 26 (October 1879):287; Kent, *Leadville*, pp. 101–5; Em-

mons, *Leadville*, p. 626; Henderson, *Mining in Colorado*, p. 89; *Grant vs. Council Bluffs Iron Works*, docket 417, FRCD.

36. Yelita to Harrison, Alpine, Colorado, October 14, 1878, Harrison Papers; *EMJ* 25 (April 1878):283; 26 (October 1878):243; 28 (October 1879):243; Kent, *Leadville*, pp. 95–97.

37. Copy of certificate of incorporation, La Plata Mining and Smelting Company, CSA; and *La Plata Mining and Smelting Company*, n.p., n.d.; *MSP* 44 (May 1882):293.

38. Kent, *Leadville*, pp. 95–97.

39. Billing to Daggett & Bristol, Salt Lake City, December 2, 1872, Bristol Family Collection; Oregon, 1:520, 527, Dun Records; *EMJ* 89 (January 1910):202.

40. See Eilers's reports, Adelberg and Raymond Collection; Rossiter W. Raymond, "Anton Eilers," pp. 762–64; Frederick E. Emrich, "Remarks at the Funeral of Anton Eilers," in *F. Anton Eilers, 1839–1917* (n.p., n.d.), p. 8; Arthur S. Dwight, "A Brief History of Blast-Furnace Lead Smelting in America," pp. 9–37; *MSP* 45 (July 1882):54.

41. *EMJ* 28 (October 1879):278; 27 (May 1879):375; Kent, *Leadville*, pp. 97–98; Henderson, *Mining in Colorado*, p. 89.

42. Kent, *Leadville* p. 112; Henderson, *Mining in Colorado*, p. 89.

43. Copies of legal documents pertaining to incorporation, CSA.

44. "Otto H. Hahn," p. 362; reports in the Adelberg and Raymond Collection; Ellsworth Daggett to Eugene Stuart Bristol, Corinne, Utah, April 13, 1871, Bristol Family Collection; Otto H. Hahn, Anton Eilers, and Rossiter W. Raymond, "The Smelting of Argentiferous Lead Ores," pp. 91–131; Otto H. Hahn, "A Campaign in Railroad District, Nevada," pp. 329–32; Dwight, "Blast-Furnace Lead Smelting," pp. 9–37.

45. Kent, *Leadville*, p. 98; Henderson, *Mining in Colorado*, p. 89; copy of certificate of incorporation, American Mining and Smelting Company, CSA.

46. Copy of certificate of incorporation, Elgin Mining and Smelting Company, CSA; prospectus, *Lake City Mining and Smelting Company*; Kent, *Leadville*, pp. 100–101; Henderson, *Mining in Colorado*, p. 89; *EMJ* 28 (October 1879):278.

47. Emmons, *Leadville*, p. 626; Kent, *Leadville*, p. 101; *EMJ* 28 (October 1879):278.

48. Kent, *Leadville*, p. 112.

49. Ibid., p. 116; Emmons, *Leadville*, pp. 641–43.

50. Emmons, *Leadville*, pp. 641–43; Kent, *Leadville*, p. 116.

51. Kent, *Leadville*, pp. 94–107, 116; *EMJ* 28 (November 1879):319;

Ernest Ingersoll, *Crest of the Continent: A Record of a Summer's Ramble in the Rocky Mountains and Beyond*, pp. 206–8.

52. Emmons, *Leadville*, pp. 627–28.

53. A. Middlebrook to Harrison, Denver, June 21, 1877, Harrison Papers; Jonathan B. Maude to Excelsior Mining and Smelting Company, Leadville, March 20, 1879; Berdell Witherell & Company, Leadville, March 20, 1879; J. B. Grant & Company to whom it may concern, Leadville, March 20, 1879, box 16, Hague Papers, HEH.

54. Emmons, *Leadville*, pp. 627–28; Holt to J. V. Farwell, Leadville, April 5, 1879, and ore receipts from J. B. Grant & Company, George Hubbard Holt Collection, CU.

55. Emmons, *Leadville*, pp. 627–29; Kent, *Leadville*, pp. 116–20.

56. Emmons, *Leadville*, pp. 639, 668–69; Kent, *Leadville*, p. 112.

57. Kent, *Leadville*, pp. 98–106; *EMJ* 28 (October 1879):278; 29 (January–June 1880):37, 211, 348; 30 (October 1880):245; 30 (November 1880):390; 31 (January 1881):43; 31 (May 1881):343.

58. John V. Farwell to Theodore Kelly, n.p., May 23, 1878; D. R. Holt to George H. Holt, Chicago, March 25, 1879; G. H. Holt to Farwell, Leadville, April 5, 1879; G. H. Holt to the executive committee, Leadville, n.d., 1879, Holt Collection.

59. Henry Howland to George H. Holt, Webster, Colorado, April 5, 1879; D. R. Holt to G. H. Holt, Chicago, April 25, July 9, 1879; Peter O. Anderson to G. H. Holt, Leadville, July 20, 1879; V. T. Axtell to G. H. Holt, Leadville, July 30, 1879; G. H. Holt to D. R. Holt, Leadville, August 26, 1879, Holt Collection; *EMJ* 29 (March 1880):211; Emmons, *Leadville*, p. 627.

60. Emmons, *Leadville*, p. 626.

61. Ibid., passim.

62. Rodman Wilson Paul, *Mining Frontiers of the Far West, 1848–1880*, pp. 130–32.

63. King to Emmons, Washington, April 20, 1880, King Papers; Emmons to Becker, Leadville, May 13, 1880, box 15, Becker Papers.

64. Emmons to Becker, Leadville, May 13, 24, 26, 29, 1880, and Denver, August 26, December 4, 1880, box 15, Becker Papers.

65. Ibid., Denver, August 26, December 4, 1880; Emmons, *Leadville*, passim.

66. Emmons to Becker, Washington, June 16, 1882, and Denver, October 31, 1882, box 15, Becker Papers; Antony Guyard to Emmons, n.p., August 13, 1882; Stanislaus Guyard to Emmons, Paris, June 9, 1884, box 11, Emmons Collection.

Chapter 5

1. Charles W. Henderson, *Mining in Colorado*, pp. 89–91, 136–44, 176; U.S. Geological Survey, *Mineral Resources of the United States, Calendar Year 1885,* pp. 200–201; U.S. Department of the Treasury, *Report of the Director of the Mint* (1881), pp. 135–39; *EMJ* 29 and 30 (1880), passim.

2. *Grant vs. Council Bluffs Iron Works*, docket 417, FRCD.

3. *EMJ* 28 (October 1879):278; 29 (1880):122, 154, 164; 30 (1880):140, 149, 357, 405; 31 (February 1881):85.

4. *EMJ* 30 (1880):140, 272, 302, 307; 33 (May 1882):236; O. J. Frost, "The Grant Lead-Smelting Works," pp. 163–64; L. A. Kent, *Leadville. The City. Mines and Bullion Product. Personal Histories of Prominent Citizens. Facts and Figures Never Before Given to the Public,* p. 104.

5. Henderson, *Mining in Colorado*, p. 176; Kent, *Leadville*, p. 112; U.S. Department of the Treasury, *Report of the Director of the Mint* (1881), p. 135; (1882), p. 416; (1883), pp. 503–4.

6. *Keyes and Arents vs. Grant and Grant*, docket 613, FRCD; *MSP* 45 (July 1882):6.

7. Dexter Papers, passim; Frank Fossett, *Colorado. Its Gold and Silver Mines, Farms and Stock Ranges, and Health and Pleasure Resorts. Tourist's Guide to the Rocky Mountains,* pp. 436, 459; *EMJ* 35 (May 1883):305; U.S. Department of the Treasury, *Report of the Director of the Mint* (1881), pp. 406, 444; (1882), pp. 486, 492–93.

8. *EMJ* 30 (June 1882):293; *MSP* 44 (June 1882):395; U.S. Department of the Treasury, *Report of the Director of the Mint* (1882), p. 500.

9. *EMJ* 30 (June 1882):293; 34 (July 1882):21; U.S. Department of the Treasury, *Report of the Director of the Mint* (1882), p. 504.

10. *EMJ* 29 (May 1880):312; 30 (1880):140, 191, 245; 31 (1881):42, 188, 246–47, 343; Henderson, *Mining in Colorado*, p. 176; U.S. Department of the Treasury, *Report of the Director of the Mint,* (1880), p. 135; (1882), p. 504.

11. *EMJ* 30 (September 1880):191; Anton Eilers to the editor, Leadville, March 24, 1881, *EMJ* 31 (April 1881):246–47; Raymond to the stockholders of the Chrysolite Silver Mining Company, New York, November 1, 1880, box 15, Hague Papers, HEH.

12. Certificate of incorporation, Arkansas Valley Smelting Company, CSA; *Arkansas Valley Smelting Company vs. Belden Mining Company*, docket 1152, FRCD.

13. Certificate of incorporation, Arkansas Valley Smelting Company, CSA.

14. U.S. Department of the Treasury, *Report of the Director of the Mint* (1882), p. 504; *EMJ* 35 (June 1883):320.

15. *EMJ*, 35 (June 1883):320, 353; 36 (1883):32, 54, 119, 219, 264, 267, 387; U.S. Department of the Treasury, *Report of the Director of the Mint* (1883), p. 366.

16. *EMJ* 31 (January 1881):24; 31 (March 1881):165; 32 (December 1881):388.

17. *EMJ* 29 (April 1880):239; 30 (October 1880):272; Hamilton S. Wicks to the editor, Leadville, November 13, 1880; *EMJ* 30(November 1880):330–31.

18. Wicks to the editor, Leadville, November 13, 1880; *EMJ* 30 (November 1880):330–31; *Thomas C. McGowan vs. the La Plata Mining and Smelting Company*, docket 667, FRCD.

19. Kent, *Leadville*, p. 112; U.S. Department of the Treasury, *Report of the Director of the Mint* (1880), p. 135; (1882), p. 504; *EMJ* 36 (October 1883):222.

20. U.S. Department of the Treasury, *Report of the Director of the Mint* (1881), pp. 414–17; Henderson, *Mining in Colorado*, pp. 137–38; *EMJ* 32 (September 1881):156.

21. U.S. Department of the Treasury, *Report of the Director of the Mint* (1882), p. 504; (1883), p. 366; *EMJ* 35 (May 1883):288, 305; 36 (1883):31, 71, 134, 250, 267, 283; *MSP* 47 (July 1883):35.

22. Henderson, *Mining in Colorado*, pp. 136–48; *MSP* 49 (July 1884):9; 51 (August 1885):132.

23. *EMJ* 31 (January 1881):43; 36 (October 1883):264; 37 (April 1884):249.

24. *EMJ* 36 (November 1883):318, 321; 36 (December 1883):365; *MSP* 53 (October 1886):269.

25. *EMJ* 37 (February 1884):89; 37 (June 1884):428; 38 (June 1884):26; 38 (July 1884):59; U.S. Department of the Treasury, *Report of the Director of the Mint* (1884), pp. 221–22.

26. *EMJ* 39 (May 1885):303.

27. *EMJ* 40 (August 1885):151; 40 (October 1885):244; R. W. Woodbury to Horace A. W. Tabor, Denver, October 14, 1885, Horace A. W. Tabor Collection, CHS.

28. *EMJ* 41 (February 1886):97; 43 (January 1887):29.

29. *EMJ* 35 (May 1883):288; 35 (June 1883):321; Morton, Bliss & Company to Hague, New York, July 24, 1882, M-4, Hague Papers, HEH.

30. *EMJ* 36 (1883):186, 264, 267, 345, 373; 37 (January 1884):9; U.S. Department of the Treasury, *Report of the Director of the Mint* (1883), p. 366.

31. *EMJ* 37 (1884):9, 412, 487; 38 (1884):8, 235, 256; 40 (October 1885):272.

32. *EMJ* 39 (1885):303, 376, 392.

33. *EMJ* 40 (1885):10, 82, 98, 151, 221, 272, 294; 41 (February 1886):136; 41 (April 1886):342; *MSP* 52 (May 1886):329; 52 (June 1886):411, 413.

34. *EMJ* 41 (June 1886):468; 42 (1886):48, 119, 154, 244, 262, 370, 424, 460; 43 (January 1887):64; 43 (May 1887):388.

35. *EMJ* 44 (July 1887):11, 29.

36. *EMJ* 35 (May 1883):305; 36 (1883):31, 71, 134, 250, 267, 283; 37 (April 1884):260.

37. *EMJ* 40 (1885):9, 82, 116; 42 (October 1886):244; 43 (January 1887):29.

38. *EMJ* 41 (February 1886):97; 43 (January 1887):29. U.S. Department of the Treasury, *Report of the Director of the Mint* (1884), p. 221.

39. Kent, *Leadville*, p. 112; U.S. Department of the Treasury, *Report of the Director of the Mint* (1880), p. 135.

40. U.S. Department of the Treasury, *Report of the Director of the Mint* (1881), pp. 414–16; (1882), p. 504; (1883), p. 366; *EMJ* 36 (July 1883):31.

41. Illinois, 41:366, and 46:8, Dun Records; *EMJ* 37 (March 1884):186; U.S. Department of the Treasury, *Report of the Director of the Mint* (1884), p. 288.

42. *EMJ* 36 (October 1883):264; 37 (June 1884):428.

43. *EMJ* 38 (1884):26, 44, 131–32, 432; U.S. Department of the Treasury, *Report of the Director of the Mint* (1884), pp. 221, 228.

44. *EMJ* 39 (1885):303, 376, 446; 40 (1885):28, 169, 287; 41 (February 1886):97.

45. *EMJ* 42 (1886):127, 243, 299, 352; 43 (January 1887):29; *MSP* 55 (March 1888):173.

46. *EMJ* 53 (April 1892):384; 53 (May 1892):551; 54 (July 1892):84–85; records of the Saint Louis Smelting and Refining Company, CSA.

47. *EMJ* 47 (January 1889):89.

48. Meyer to Charles Merriam, New York, December 19, 1888; Merriam to Higginson, Boston, December 21, 1888; James Jackson to

Higginson, Boston, May 13, 1890; Kuhn, Loeb & Company to Higginson, New York, May 15, 1890, Higginson Papers; Committee on Stock List, New York Stock Exchange, *Consolidated Kansas City Smelting and Refining Company.*

49. Committee on Stock List, *Consolidated Kansas City Company*; *EMJ* 47 (February 1889):192; 53 (April 1892):384.

50. *EMJ* 43 (1887) through 55 (1893), passim.

Chapter 6

1. Robert G. Athearn, *Union Pacific Country*, pp. 132–33, 253–55, 259–61; Robert G. Athearn, *Rebel of the Rockies: A History of the Denver and Rio Grande Western Railroad*, pp. 15, 25, 43. The Union Pacific took control of the Kansas Pacific in 1879.

2. Nathaniel P. Hill to Henry M. Teller, Black Hawk, October 10, 1877, Teller Collection, DPL; *U.S. vs. N. P. Hill et al.*, docket 121, FRCD; Rickard, "Richard Pearce," p. 408.

3. Traxler, "Colorado Central," pp. 70–71.

4. *Rocky Mountain News*, April 3, 1878.

5. *Denver Daily Times*, February 2, 1878; Hill to Augustine G. Langford, Black Hawk, March 28 and April 10, 1877, Langford Collection, microfilm 1023, University of Colorado Library.

6. Certificate of incorporation, Denver and Rocky Mountain Railway Company, book A, pp. 481–85, CSA; *Daily Times*, January 14, 1878.

7. *Daily Times*, February 2, 11, 15, 20, March 13, 19, June 27, 1878; *Rocky Mountain News*, May 30, 1878.

8. *Daily Times*, February 1, 2, March 19, 29, April 1, May 7, 8, 1878; *Rocky Mountain News*, May 23, August 1, September 17, 1878; January 1, 1879; *Denver Tribune*, January 1, 1879, in Dawson scrapbook 36, p. 417; Frank Fossett, *Colorado. Its Gold and Silver Mines, Farms and Stock Ranges, and Health and Pleasure Resorts. Tourist's Guide to the Rocky Mountains*, pp. 240–42.

9. Wolcott to Hague, Denver, May 18, 1879, box 15, Hague Papers, HEH; *Denver Tribune*, January 1, 1879; *Rocky Mountain News*, January 1, 1879; "Value of the Gold, Silver, and Copper . . . , box 37, C. Hill Papers, CHS.

10. Certificates of condition, Boston and Colorado Smelting Company, 1878–80, CM.

11. Chaffee to W. H. Pierce, New York, May 30, 1878, in the *Daily*

Times, June 7, 10, 1878; *Rocky Mountain News*, December 25, 1878.

12. *Rocky Mountain News*, November 5, December 27, 1878, and other issues.

13. Ibid., January 10, 1879. The quotation is in *EMJ* 27 (January 1879):55. See also Colorado General Assembly, *Senate Journal of the General Assembly of the State of Colorado, Second Session* (1879), pp. 107, 115; *House Journal of the General Assembly of the State of Colorado, Second Session* (1879), p. 112; *MSP* 38 (February 1879): 92.

14. "Value of the Gold, Silver, and Copper . . . ," box 37, C. Hill Collection, CHS; *EMJ* 31 (January 1881):31; 35 (January 1883):7; *Rocky Mountain News*, January 1, December 31, 1884; *Denver Republican*, January 1, 1882.

15. T. A. Rickard, "Richard Pearce," p. 408.

16. Rossiter W. Raymond, "Henry Williams," p. 82.

17. Stock certificate no. 1, Colorado and Montana Smelting Company, box 1, C. Hill Collection, CHS; Patrick Henry McLatchy, "A Collection of Data on Foreign Corporate Activity in the Territory of Montana, 1864–1889," pp. 91–92; Rickard, "Richard Pearce," p. 408.

18. Raymond, "Henry Williams," p. 82; U.S. Department of the Treasury, *Report of the Director of the Mint* (1882), p. 232; (1883), p. 250.

19. Certificate of incorporation, Miner's Smelting and Reduction Company, CSA; *Rocky Mountain News*, November 10, 1881; notebook no. 13, (1884), pp. 121–22, 138–39, Weitbrec Collection. O'Niel described the process to a man identified only as "McN" (probably McNair), who in turn divulged the details to Weitbrec. The description is similar to that given by Harold V. Pearce in his article "The Pearce Gold-Separation Process."

20. *Rocky Mountain News*, November 10, 1881; March 8, 1882; November 18, 1883; *EMJ* 34 (September 1882):124; 35 (March 1883):143; 36 (October 1883):259. The quotation is in the last reference.

21. Notebook no. 8, pp. 183–84, Howe Collection; Frost, "Grant Lead-Smelting Works," pp. 163–64.

22. O. J. Frost, "The Grant Lead-Smelting Works," pp. 163–64; *EMJ* 34 (1882):21–22, 61, 151, 203; Malvern W. Iles to Thomas Egleston, Denver, March 5, 1884, Egleston Collection; *MSP* 45 (July 1882):21; 45 (October 1882):281; 46 (April 1883):269.

23. Duane A. Smith, *Horace Tabor: His Life and the Legend*, pp. 212–25; and *Rocky Mountain News*, August–November 1882, passim.

24. Certificate of incorporation, Omaha and Grant Smelting and

Refining Company, CSA; *EMJ* 36 (August 1883):87; *MSP* 47 (August 1883):101.

25. *EMJ* 36 (August 1883):102; 37 (March 1884):241; 37 (April 1884):317; 40 (July 1885):63; *MSP* 48 (March 1884):209.

26. *EMJ* 36 (1883):235, 329, 345; certificate of incorporation, Fryer Hill Smelting Company, CSA; U.S. Department of the Treasury, *Report of the Director of the Mint* (1882), p. 504; (1883), p. 366; (1884), p. 221.

27. *EMJ* 37 (1884):240, 317, 412; 38 (July 1884):44; 39 (1885):127, 303, 322; 40 (1885):9, 28, 61, 63, 244, 391, 406; 41 (February 1886):116; *MSP* 48 (March 1884):209.

28. *EMJ* 40 (July 1885):28, 61; *MSP* 51 (July 1885):73.

29. *EMJ* 41 (1886):216, 289, 396; 42 (1886):172, 262, 442; 43 (January 1887):28.

30. Certificate of incorporation, Holden Smelting Company, CSA.

31. *EMJ* 43 (February 1887):100; 43 (June 1887):425; 44 (December 1887):427.

32. Certificate of incorporation, Globe Smelting and Refining Company, CSA; Harvey O'Connor, *The Guggenheims: The Making of an American Dynasty*, pp. 72–73.

33. Dennis Sheedy, *Autobiography of Dennis Sheedy*, pp. 48–54; Kedro, "Czechs and Slovaks," passim.

34. Certificate of incorporation and other records, Globe Smelting and Refining Company, CSA; O'Connor, *Guggenheims*, pp. 72–73.

35. *EMJ* 48 (August 1889):133; John M. Walker to John F. Campion, Denver, October 12, 1892, box 3, and letter fragment about 1890, box 7, Campion Papers; W. E. Newberry to Holden Smelting Company, Aspen, March 14, 1888, box 1, Brown Papers; Charles W. Henderson, *Mining in Colorado*, pp. 90–93.

36. Henry Raup Wagner, *Bullion to Books*, passim.

37. *EMJ* 54 (November 1892):444; George D. Rogers to Globe Smelting and Refining Company, New York, December 22, 1898, February 11, 23, 1899, Record Group 1, Rockefeller Papers.

38. *EMJ* 43 (January 1887):28; 43 (June 1887):425; 44 (July 1887):83; 44 (October 1887):317.

39. *EMJ* 45 (January 1888):400; Henderson, *Mining in Colorado*, pp. 91–93; *MSP* 55 (August 1887):105.

40. James H. Devereux to Omaha and Grant Smelting and Refining Company, Aspen, October 8, 22, 29, 1885; W. B. Devereux to Omaha and Grant Smelting and Refining Company, Aspen, March 15, October

22, 1887; invoices sent to the Omaha and Grant enterprise, Brown Papers; W. Byrd Page to John F. Campion, Denver, November 9, 1892; Grant to Campion, Denver, May 31, 1893, box 3, Campion Papers; Henderson, *Mining in Colorado*, passim.

41. Henry E. Wood, "I Remember," Wood Collection.

42. George C. Clark to Higginson, New York, December 26, 1891, Higginson Collection.

43. Clark, Dodge & Company, to Higginson, New York, November 12, 1891, Higginson Collection; *MSP* 62 (January 1891):53.

44. Gardner M. Lane to Higginson, London, November 7, 11, 1891; Clark, Dodge & Company to Higginson, New York, November 12, 1891; D. G. Clark to Higginson, n.p., November 30, 1891; George C. Clark to Higginson, New York, December 26, 1891; A. A. H. Boissevain to Lee, Higginson & Company, New York, December 28, 1891, Higginson Collection.

45. Clark, Dodge & Company to Higginson, New York, November 12, 1891, Higginson Collection; certificate of incorporation, Omaha and Grant Smelting Company, CSA; *EMJ* 55 (February 1893):104.

46. Hill as quoted in the *Rocky Mountain News*, January 31, 1885; Hale, "First Successful Smelter in Colorado," p. 167.

47. Certificates of condition, Boston and Colorado Smelting Company, 1884–90, CM.

48. Edward D. Peters, Jr., *Modern Copper Smelting*, pp. 444–47; *MSP* 45 (December 1882):385; 62 (January 1891):53.

49. Peters, "Reverberatory Practice at Argo," p. 190.

50. Peters, *Modern Copper Smelting*, pp. 200–214; H. O. Hofman, *Metallurgy of Copper*, pp. 133–35; "Pearce Turret Furnace," p. 513; Crawford Hill to Jesse D. Hale, Denver, July 16, 1901, Hill Collection, DPL; Crawford Hill to F. W. Howbert, Denver, January 25, 1910, box 1, C. Hill Collection, CHS.

51. Pearce, "Progress of Metallurgical Science in the West," pp. 55–72; *MSP* 48 (June 1884):399.

52. *EMJ* 44 (November 1887):381; 46 (September 1888):179; 47 (June 1889):528; Hale, "First Successful Smelter in Colorado," p. 165; *MSP* 46 (March 1883):270.

53. Fred Bulkley to Boston and Colorado Smelting Company, Aspen, January 31, 1888; Fred Bulkley to Hill, Aspen, February 8, 1888, Brown Papers; Pearce, "Progress of Metallurgical Science in the West," p. 58; Hale, "First Successful Smelter in Colorado," p. 165; *EMJ* 43 (1887) to 57 (1894), passim.

54. "Value of the Gold, Silver, and Copper . . . ," box 37, C. Hill Collection, CHS; Henderson, *Mining in Colorado*, pp. 109, 122.

Chapter 7

1. Julius Grodinsky, *Transcontinental Railway Strategy, 1869–1893: A Study of Businessmen*, passim; Robert G. Athearn, *Rebel of the Rockies: A History of the Denver and Rio Grande Western Railroad*, passim; H. Lee Scamehorn, *Pioneer Steelmaker in the West: The Colorado Fuel and Iron Company, 1872–1903*, passim.

2. Interview with Alfred W. Geist, Bancroft Collection; *Keyes and Arents vs. Grant and Grant*, docket 613, FRCD; William Geist to George Jarvis Brush, Florence, Italy, August 6, 1869, Brush Collection.

3. *Colorado Daily Chieftan*, June 8, 16, August 16, September 10, 12, 1878; copy of letter in letterbook 9, p. 798, Olcott Papers.

4. Emmons to Becker, Denver, May 12, 1880, box 15, Becker Papers; *Colorado Daily Chieftan*, October 17, December 21, 1878, April 3, 1879.

5. Notebook no. 8, p. 185 and passim, Howe Collection; *EMJ* 32 (July 1881):38; *Colorado Daily Chieftan*, August 14, October 18, 1880, April 20, May 11, 1881.

6. Geist interview, Bancroft Collection; *Pueblo Daily Chieftan*, December 30, 1881, April 2, September 29, 1882.

7. Mahlon D. Thatcher to J. L. Ward, Pueblo, February 27, 1880; Ward to Nickerson, Louisville, Kentucky, March 1, 2, 8, 27, April 1, 22, 1880, RR 309, Atchison, Topeka, and Santa Fe Railroad Collection; Nickerson to Hague, Boston, March 25, 1879, M-3, Hague Papers, HEH; Emmons to Becker, Denver, May 12, 1880, box 15, Becker Papers; *EMJ* 34 (November 1883): 259.

8. *EMJ* 37 (January 1884):71; 38 (October 1884):235; 39 (May 1885):303; 42 (August 1886):137; *MSP* 48 (March 1884):227; 51 (September 1885):193; 54 (March 1887):193.

9. *EMJ* 43 (March 1887):227; 44 (November 1887):401; 47 (March 1889):284; 47 (May 1889):485; 50 (December 1890):723; D. A. Wightman Collection.

10. *EMJ* 43 (1887) through 55 (1893), passim.

11. " 'Mexico' or 'Mexican Town,' " pp. 245–50.

12. Emmons to Higginson, Denver, December 9, 1882, Higginson Collection; Raymond, "Anton Eilers," pp. 762–64.

13. *EMJ* 35 (March 1883):181; 36 (September 1883):135; 36 (October 1883):264; "Otto H. Hahn," p. 362.

14. Raymond, "Anton Eilers," pp. 762–64; *EMJ* 37 (April 1884):261; *MSP* 51 (July 1885):73.

15. Rossiter Raymond to Sarah Raymond, Pueblo, November 12, 1887, Author's Papers; Arthur S. Dwight, "Reminiscences," in Thomas A. Rickard, ed. *Rossiter Worthington Raymond: A Memorial*, p. 67; Rossiter W. Raymond, "Anton Eilers," pp. 762–64.

16. *EMJ* 46 (November 1888):398; 48 (July 1889):13; Dwight to Campion, box 3, Campion Papers.

17. Hewitt to Eilers, New York, May 11, 1888, and Hewitt to W. S. Gurnee, New York, May 15, 1888, Abram S. Hewitt Papers; Henry Seligman to Albert Seligman, New York, August 5, November 7, 1887, Seligman Collection; Rossiter W. Raymond to Sarah Raymond, Smelter, Montana, November 9, 12, 1889, Author's Papers; Hakola, "Samuel T. Hauser," p. 290.

18. *EMJ* 37 (February 1884):90; *Pueblo Daily Chieftan*, January 26, February 13, 1884.

19. *EMJ* 37 (April 1884):261; 38 (October 1884):271; 39 (April 1885):267; 47 (March 1889):241; *Pueblo Daily Chieftan*, December 4, 5, 1885; *MSP* 49 (July 1884):9.

20. *EMJ* 25 (1878) to 47 (1889), passim.

21. New York, 405A:1162, and 248:2483, Dun Records; Isaac F. Marcosson, *Metal Magic: The Story of the American Smelting and Refining Company*, pp. 22–29.

22. New York, 259:3594–95, Dun Records.

23. Harvey O'Connor, *The Guggenheims: Making of an American Dynasty*, pp. 42–55; U.S. Department of the Treasury, *Report of the Director of the Mint* (1881), pp. 498–99; (1883), pp. 358–59.

24. Charles W. Henderson, *Mining in Colorado*, pp. 139–48; O'Connor, *Guggenheims*, pp. 54–55.

25. *EMJ* 45 (January 1888):58.

26. O'Connor, *Guggenheims*, pp. 72–79.

27. Certificate of incorporation, Denver Smelting and Refining Company, CSA.

28. *EMJ* 45 (January 1888):58.

29. O'Connor, *Guggenheims*, pp. 78–80; Hoyt, *Guggenheims*, p. 65.

30. *EMJ* 46 (September 1888):180; 46 (October 1888):331.

31. *EMJ* 46 (December 1888):465; 47 (1889):72, 192–93, 218–19, 241; Benjamin Guggenheim to the editor, Pueblo, March 4, 1889, in *EMJ* 47 (March 1889):253; *MSP* 59 (August 1889):102.

32. Records of the Philadelphia Smelting and Refining Company, CSA; O'Connor, *Guggenheims*, pp. 80–83.

33. O'Connor, *Guggenheims*, pp. 82–83 and passim; *EMJ* 48 (July 1889):13; 48 (September 1889):252.

34. Copy of letter, letterbook 9, p. 798, Olcott Papers; *EMJ* 17 (1874) to 30 (1880), passim.

35. Athearn, *Rebel of the Rockies*, passim.

36. *EMJ* 30 (1880):28, 111, 249; 31 (February 1881):92; New York, 414:383, Dun Records.

37. Iowa, 33:361, 374, Dun Records; *EMJ* 30 (November 1880):357; 31 (February 1881):92.

38. *EMJ* 37 (1884):9, 186, 317, 392; 39 (May 1885):377; 39 (June 1885):446; 41 (January 1886):26; 41 (March 1886):198; 42 (July 1886):48, 65; 43 (January 1887):29; *MSP* 45 (October 1882):278.

39. *EMJ* 29 (1880), passim.

40. San Juan Smelting and Mining Company, *Annual Report, 1888*, Bell Papers.

41. Bell to Sheldon [?], n.p., n.d.; Porter to Bell, Durango, October 20, 1890, and Denver, November 14, 1890, Bell Papers; *EMJ* 45 (April 1888):311.

42. San Juan Smelting and Mining Company, *Annual Reports, 1888–1890;* Porter to Bell, Denver, November 14, 15, 1890, Bell Papers; *MSP* 59 (August 1889):105.

43. Porter to Bell, Denver, November 14, 15, 26, December 4, 15, 1890, Durango, January 8, 1891, and Denver, April 1, 1891, Bell Papers.

44. Walter Hinchman to Palmer, New York, August 20, 1891, Bell Papers. Palmer made comments on the letter and forwarded it to Bell.

45. Porter to Bell, n.p., January 15, 1892, Durango, February 18, 27, March 2, 1892, and Porter to Henry Amy, Durango, February 15, 1892, Bell Papers.

46. Porter to Bell, Denver, March 26; New York, April 3; Denver, October [?], November 10, December 22, 1892; January 25, February 21, 1893; George Foster Peabody to Bell, n.p., January 27, 1893, Bell Papers.

47. Walter Renton Ingalls, *Lead and Zinc in the United States, Comprising an Economic History of the Mining and Smelting of the Metals and the Conditions which have Affected the Development of the Industries*, pp. 216–19; *MSP* 46 (January 1883):18; 48 (April 1884):286.

48. Charles W. Bennett, W. G. Van Horne, and Thomas H. Carter, *Before the Hon. Secretary of the Treasury. In the Matter of the Hearing*

on the Question of Admitting Free of Duty Foreign Ores Containing both Lead and Silver, passim; *MSP* 55 (June 1888):425; 56 (January 1888):36.

49. See Meyer's testimony in *Revision of the Tariff: Hearings before the Committee on Ways and Means, 1889–90*, pp. 1272 ff. The quotation is on p. 1272.

50. Eilers to Edward O. Wolcott, Pueblo, May 4, 1889, in *Revision of the Tariff: Hearings before the Committee on Ways and Means, 1889–90*, pp. 1283–84; Eilers to McKinley, New York, March 10, 1890, in ibid., p. 1276; see Meyer's testimony in ibid., pp. 1272–86.

51. Ingalls, *Lead and Zinc*, pp. 216–19; *MSP* 62 (January 1891):35; 62 (February 1891):83.

52. Marvin D. Bernstein, *Mexican Mining Industry, 1890–1950: A Case in the Interaction of Politics, Economics, and Technology*, pp. 39, 60–61; Towne to Higginson, New York, February 25, 1892; Meyer to Higginson, Kansas City, Missouri, May 25, 1891, New York, June 6, 20, August 15, 1891, Higginson Papers; *MSP* 62 (January 1891):35.

53. Committee on Stock List, *Consolidated Kansas City Company*; Bernstein, *Mexican Mining Industry*, pp. 37, 53.

54. Bernstein, *Mexican Mining Industry*, pp. 37–39.

55. *EMJ* 49 (March 1890):282, 284.

Chapter 8

1. Walter Renton Ingalls, *Lead and Zinc in the United States, Comprising an Economic History of the Mining and Smelting of the Metals and the Conditions which have Affected the Development of the Industries*, pp. 320–26. The quotation is in Ben Stanley Revett to John F. Campion, Denver, June 27, 1893, box 3, Campion Papers.

2. Henry E. Wood to Belle M. Wood, two letters, Denver, June 26, 1893, Wood Collection; Fred G. Bulkley to the president and board of trustees, Aspen, August 11, 1893, Brown Papers. The quotation is W. W. Allen to Franklin Ballou, in Ballou to John F. Campion, Leadville, June 29, 1893, Campion Papers.

3. *EMJ* 56 (August 1893):171, 197; 56 (September 1893):273, 351.

4. Dwight to Campion, Pueblo, August 23, 1893; L. B. Beach to Campion, Pueblo, August 31, 1893, Campion Papers; Fred G. Bulkley to Philadelphia Smelting and Refining Company, Aspen, two letters, August 3, 1893, Brown Papers.

5. James H. Devereux to Omaha and Grant Smelting and Refining Company, Aspen, October 22, 29, 1885; Fred G. Bulkley to W. W.

Hardinge, Aspen, June 14, 1892; Bulkley to Wheeler, Aspen, October 12, December 26, 1892; January 18, July 3, 1893, Brown Papers.

6. Frank Bulkley to J. E. Schwartz, Aspen, July 27, 1893. The quotation is in Bulkley to Wheeler, Aspen, July 10, 1893. See also Bulkley to Wheeler, Aspen, July 12, 1893, Brown Papers.

7. Ingalls, *Lead and Zinc*, pp. 34–35.

8. Henslee to Campion, Denver, April 20, 1894, Campion Papers; Frank Bulkley to Wheeler, Aspen, June 4, December 1, 1894, Brown Papers.

9. Ingalls, *Lead and Zinc*, passim.

10. *EMJ* 56 (1893) to 59 (1895), passim.

11. *EMJ* 57 (January 1894):36–37, 86; 57 (February 1894):181. The quotation is in the latter issue.

12. *EMJ* 58 (July 1894):85.

13. *EMJ* 61 (July 1896):61; 62 (December 1896):565; 63 (January 1897):9; 64 (November 1897):642, 645; Carl Koelle to John J. Blow, El Paso, September 11, 1894; Joseph H. Weddle to Blow, Leadville, June 6, 1896; Arkansas Valley Smelting Company to B. F. Follett, Leadville, December 20, 1896, Blow Papers; contract between the Highland Chief mine and the Arkansas Valley Smelting Company, December 28, 1895; and Weddle to Campion, Denver, December 20, 1898, Campion Papers.

14. *EMJ* 58 (July 1894):85; 63 (January 1897):73; 63 (February 1897):145; 66 (August 1898):265; Committee on Stock List, *Consolidated Kansas City Company*; H. O. Hofman, *Metallurgy of Lead and the Desilverization of Base Bullion*, pp. 173–75.

15. Committee on Stock List, *Consolidated Kansas City Company*; Meyer to Higginson, Chihuahua, Mexico, January 30, 1897, Higginson Papers.

16. Meyer to Higginson, Argentine, Kansas, May 22, 1897, Higginson Papers.

17. Edward D. Peters, Jr., *Principles of Copper Smelting*, pp. 213–338.

18. Ballou to Smith, Leadville, September 27, 1891, May 19, 1894, Smith Correspondence.

19. *EMJ* 54 (1892):159, 230, 373, 421.

20. Ballou to Smith, Leadville, May 19, June 7, 1894, Smith Correspondence; *EMJ* 57 April (1894):326; 57 (May 1894):469; 58 (September 1894):205, 300.

21. Frank Bulkley to Wheeler, Aspen, July 23, 1894, Brown Papers; Ballou to Blow, Leadville, June 22, 1896; Ballou to W. B. Page and Blow, Leadville, August 7, 1897, Blow Papers.

22. Dwight to Blow, Pueblo, September 3, 11, 17, 1896, Blow Papers; Weddle to Emmons, Denver, February 17, 1896, Emmons Collection; *EMJ* 63 (1897):9, 519, 548, 670.

23. *EMJ* 53 (1892):457, 480, 577; 54 (July 1892):13, 62; 55 (April 1893):325.

24. *EMJ* 55 (June 1893):589; 56 (July 1893):85, 119.

25. *EMJ* 57 (1894):86, 181, 230, 258; 58 (October 1894):364.

26. *EMJ* 60 (December 1895):568; 62 (July 1896):85; F. A. Keith to Blow, Leadville, May 3, 1895, Blow Papers; L. S. Smith to John F. Campion, Denver, October 10, 1896, Campion Papers.

27. *EMJ* 63 (April 1897):362; 63 (May 1897):548; 66 (August 1898):265.

28. *EMJ* 57 (April 1894):326; 58 (July 1894):85.

29. Unsigned papers and documents concerning the sale of bonds, Bell Papers.

30. *EMJ* 59 (March 1895):577; 59 (April 1895):385.

31. Ernest J. H. Amy to Denver City Mining Company, Durango, April 8, May 30, 1895, Blow Papers; "Franklin Guiterman," *EMJ* 99 (May 1915):872.

32. Guiterman to Blow, Durango, May 21, 1895, Blow Papers; *EMJ* 60 (1895):37, 353, 495.

33. Testimony of Hill in U.S. Congress, House, *Report of the Industrial Commission*, 12:282.

34. Hill to Wolcott, Denver, January 14, 1898; Wolcott to Hill, New York, February 18, 1897; C. S. Tuckerman to Hill, Boston, February 27, June 9, 1897, box 1, C. Hill Collection, CHS; Wolcott to Higginson, New York, February 14, 23, 1897, and Boston, March 1, 3, 1897, Higginson Papers; Higginson to Hague, Boston, February 10, 1897; and "Report on the Colorado Smelting and Mining Company," boxes 11, 16, Hague Papers, HEH.

35. William B. Gates Jr., *Michigan Copper and Boston Dollars: An Economic History of the Michigan Copper Mining Industry*, pp. 85–89, 249.

36. Records of the Boston and Colorado Smelting Company, CSA; testimony of Hill, in U.S. Congress, House, *Report of the Industrial Commission*, 12:372.

37. Hill to Edna [?], Denver, February 29, 1899; Hill to Charles M. Coburn, Denver, January 16, 1900; Crawford Hill to Wendell P. Hale, Denver, May 8, 1900, Hill Collection, DPL.

38. *EMJ* 55 (1893) to 69 (1900), passim.

39. Weddle to Campion, Leadville, April 27, 1899, Campion Papers.

40. Meyer to Daniel Guggenheim, London, October 6, 1897, copy in Adams Papers. The quotation is in Meyer to Higginson, Argentine, Kansas, May 22, 1897, Higginson Papers.

41. *EMJ* 64 (November 1897):602, 631; 64 (December 1897):661, 696, 705, 721.

42. Isaac F. Marcosson, *Metal Magic: The Story of the American Smelting and Refining Company*, pp. 57–65.

43. Testimony of E. R. Chapman, in U.S. Congress, House, *Report of the Industrial Commission*, 13:93–98; Luther Conant, Jr., "Report on the American Smelting and Refining Company . . . ," passim, file 3198-1, Record Group 122, Bureau of Corporations, Records of the Department of Commerce.

44. Grant to Campion, Denver, February 24, 1899; Meyer to Campion, New York, March 27, April 10, 1899, Campion Papers; William H. Brevoort to R. H. Reid, New York, February 27, March 8, April 10, 12, 29, 1899, Smith Correspondence; Committee on Unlisted Securities, *American Smelting and Refining Company.*

Chapter 9

1. Meyer to John F. Campion, Argentine, Kansas, April 28, 1899, Campion Papers.

2. Ballou to Campion, Denver, May 11, 1899; Henry Lyne to Campion, Denver, May 14, 1902; Weddle to Campion, Leadville, April 27, 1899; Julius Rodman to Ibex Mining Company, Leadville, June 24, 1902; and William B. McDonald to Campion, Leadville, July 8 and 15, 1902, Campion Papers.

3. David L. Lonsdale, "The Movement for an Eight-hour Law in Colorado, 1893–1913," p. 85.

4. U.S. Bureau of Labor Statistics, *Biennial Report, 1899–1900,* p. 170; Campion to Weddle, Leadville, May 1, 1899, Campion Papers.

5. U.S. Bureau of Labor Statistics, *Report, 1899–1900*, pp. 171–84; *EMJ* 67 (June 1899):726; 68 (July 1899):46.

6. Campion to Meyer, Leadville, June 18, 1900; Meyer to Campion, Nonquitt, Massachusetts, July 3, 1900, Campion Papers; Stockholders Committee of Investigation, *Report to the Stockholders of the American Smelting and Refining Company*, p. 87.

7. American Smelting and Refining Company, Annual Report, 1900; affidavit of Edward W. Nash in *William M. Donald et al. vs. the American Smelting and Refining Company*, file 3198–9, Record Group 122, NA.

8. Towne to Higginson, New York, June 25, 1900, copy in Peabody Collection; George Foster Peabody to Higginson, New York, November 17, 1899, Higginson Papers.

9. *William M. Donald et al. vs. ASARCO*, file 3198–9, Record Group 122, NA.

10. See entries in letterbook 23, pp. 270–76, series 2, in the Whitney Collection; Isaac F. Marcosson, *Metal Magic: The Story of the American Smelting and Refining Company*, p. 63.

11. Conant, "Report on American Smelting and Refining Company," pp. 11–16; *William M. Donald et al. vs. ASARCO*, file 3198–9, Record Group 122, NA.

12. Marcosson, *Metal Magic*, pp. 301–2; Weddle to Campion, New York, February 28, 1902, Campion Papers.

13. American Smelting and Refining Company, *Annual Report, 1903*, p. 4.

14. Charles W. Henderson, *Mining in Colorado*, pp. 146–60.

15. Certificate of incorporation, United States Zinc Company, CSA; *EMJ* 72 (October 1901):437.

16. *EMJ* 73 (March 1902):422; 74 (November 1902):596; 75 (1903):13, 680, 942. The quotation is in *EMJ* 81 (March 1906):632.

17. O. Pufahl, "The Works of the United States Zinc Company at Pueblo, Colorado," p. 1232.

18. *EMJ* 82 (December 1906):1186; 88 (August 1909):382; Henderson, *Mining in Colorado*, pp. 153–66 and passim.

19. American Smelting and Refining Company, *Annual Report, 1903*, p. 7; *EMJ* 84 (July 1907):126; Kenneth S. Guiterman, "Mining Coal in Southern Colorado," pp. 1009–15.

20. Interview with Darwin J. and Mary Pope; Andrew E. Beer, New York, July 11, 1974, *EMJ* 74 (August 1902):257; 81 (May 1906):985; Stockholders Committee of Investigation, Report to the Stockholders of the American Smelting and Refining Company, p. 87.

21. *EMJ* 67 (1899) through 82 (1906), passim. The quotation is in Claude L. McKesson to Theodore Roosevelt, Black Hawk, February 16, 1906, file 3198–4, Record Group 122, NA.

22. *EMJ* 75 (June 1903):978.

23. Lonsdale, "Movement for an Eight-hour Law," pp. 207–14; *EMJ* 75 and 76 (1903), passim.

24. *EMJ* 75 (1901) through 76 (1903), passim.

25. John Fahey, *Ballyhoo Bonanza: Charles Sweeny and the Idaho Mines*, pp. 174–78.

26. Rockefeller, Jr., to Gates, New York, May 8, 1903; Rockefeller,

Jr., to Rockefeller, New York, November 4, 1903, Rockefeller Papers.

27. Rockefeller, Jr. to Rockefeller, New York, September 15, 1902, Rockefeller Papers.

28. Rockefeller, Jr., to Gates, New York, May 8, August 5, 12, 1903, Rockefeller Papers; Fahey, *Ballyhoo Bonanza*, pp. 182, 189.

29. Rockefeller, Jr., to Rockefeller, New York, August 5, September 4, 30, 1903, Rockefeller Papers.

30. Gates to Rockefeller, New York, February 7, March 13, 15, May 18, 1905; Gates to Charles O. Heydt, New York, April 5, 1905; Frederick J. Lovatt to Rockefeller, New York, April 28, 1905, Rockefeller Papers.

31. Stockholders Committee of Investigation, *Report to the Stockholders of the American Smelting and Refining Company*, p. 8.

32. Fahey, *Ballyhoo Bonanza*, pp. 182, 189; American Smelting and Refining Company, *Annual Report, 1905*, pp. 8–11, and *Annual Report, 1906*, pp. 7–9.

33. *EMJ* 70 (September 1900):377; 70 (December 1900):767; 71 (January 1901):14; 72 (1901):407, 579, 645.

34. Kenneth L. Fahnestock to Campion, Leadville, August 5, 1902; Campion to E. C. Simmons, Denver, September 25, 1902, Campion Papers; *EMJ* 73 (April 1902):595; 73 (June 1902):872; 74 (1902):59, 458, 660, 827.

35. Campion to E. C. Simmons, Leadville, September 22, 1903, Campion Papers; *EMJ* 76 (1903), passim.

36. Henry Lyne to Fahnestock, Denver, January 6, 1905; McDonald to Fahnestock, Leadville, May 2, 15, 1905. The quotation is in Guiterman to Campion, Denver, June 3, 1905, Campion Papers; *EMJ* 79(May 1905):885.

37. McDonald to Fahnestock, Leadville, December 6, 1905, Campion Papers; American Smelting and Refining Company, *Annual Report, 1906*.

38. *EMJ* 82 (July 1906):83; 82 (September 1906):562; Guiterman to Campion, Denver, May 23, June 14, 1906. The quotation is in the former letter; Campion to Smith, Denver, September 4, 1906, Campion Papers.

39. Guiterman to Campion, Denver, January 5, 1907; Henry Lyne to Campion, Denver, January 5, 1907; McDonald to Ibex Mining Company, Leadville, January 5, February 20, 1907, Campion Papers; *EMJ* 83 (May 1907):1021.

40. Henderson, *Mining in Colorado*, pp. 160–77.

41. O. Pufahl, "The Globe Plant of the American Smelting and Refining Company," pp. 1009–15.

42. Henderson, *Mining in Colorado*, passim.

43. For statistical information on ASARCO's plants in Colorado, see various documents from the auditor's office in the Roeser Collection; Pufahl, "Pueblo Lead Smelters," p. 889; *EMJ* 85 (February 1908):382; 87 (February 1909):419; 93 (June 1912):1149; 95 (April 1913):778, 826.

44. Collins, "Colorado," pp. 97–99. The quotation is in Guiterman to Campion, Denver, July 30, 1908, Campion Papers.

45. Guiterman, "Status of Mining and Smelting in Colorado," pp. 1009–10.

Chapter 10

1. Hill to George D. Edmands, Denver, November 5, 17, 1900, Hill Collection, DPL.

2. Hill to Alice Hill, Denver, November 8, 11, 1901, Hill Collection, DPL.

3. The initial quotation is in Hill to Isabel Hill, Denver, October 29, 1901, the second in Crawford Hill to Costello Converse, Denver, November 11, 1901. See also Hill to Converse, Denver, December 13, 1901, Hill Collection, DPL.

4. T. A. Rickard, "Richard Pearce," p. 408; *EMJ* 77 (February 1904):211.

5. Hill to Converse, Denver, October 31, November 12, December 11, 1901, January 2, 1902, Hill Collection, DPL.

6. Hill to Converse, Denver, November 11, December 13, 1901, Hill Collection, DPL; H. V. Pearce, "Improved Method of Slag-Treatment at Argo," p. 89.

7. Hill to Converse, Denver, October 31, 1901; Hill to White, Denver, October 31, December 31, 1901, Hill Collection, DPL; William B. Gates, Jr., *Michigan Copper and Boston Dollars: An Economic History of the Michigan Copper Mining Industry*, pp. 85–89; "Value of the Gold, Silver, and Copper . . . ," box 37, C. Hill Collection, CHS.

8. Hill to Converse, Denver, November 12, 1901, Hill Collection, DPL.

9. Pearce, "Improved Method of Slag-Treatment at Argo," pp. 28–29; Hale, "First Successful Smelter in Colorado," pp. 161–67.

10. *EMJ* 75 (March 1903):423; 75 (April 1903):642; 78 (July 1904):33; 79 (January 1905):19, 81; 81 (February 1906):301; Charles W. Henderson, *Mining in Colorado*, pp. 109, 126.

11. Pearce, "Improved Method of Slag-Treatment at Argo," p. 183.

12. Converse to the stockholders, Denver, September 20, 1909, box

37, C. Hill Collection, CHS; Hale, "First Successful Smelter in Colorado," p. 167.

13. H. V. Pearce, "The Pearce Gold- Separation Process," pp. 722–34.

14. White to Hill, Boston, August 24, 1909, box 3, C. Hill Collection, CHS; Converse to the stockholders, Denver, September 20, 1909, box 37, C. Hill Collection, CHS.

15. White to Hill, Boston, October 27, November 1, 1909, box 3, C. Hill Collection, CHS.

16. *Minutes of a Special Meeting of the Stockholders of the Boston and Colorado Smelting Company: Called to Consider the Question of Dissolution; Notice of Dissolution of the Boston and Colorado Smelting Company*; Hill to F. W. Howbert, Denver, January 28, 1910, box 37, C. Hill Collection, CHS; Henderson, *Mining in Colorado*, p. 15.

17. For a plethora of letters and documents pertaining to sales of the firm's assets and property, see box 37, C. Hill Collection, CHS.

18. Certificate of incorporation, Clear Creek Mining and Reduction Company, CSA.

19. Hague to Edward H. Harriman, New York, May 4, 1899, box 16, Hague Papers, HEH.

20. *EMJ* 71 (January 1901):187, 253; 72 (August 1901):147; 73 (1902):287, 327, 496, 905; 74 (1902):420, 422, 596, 762; 75 (June 1903):942; 76 (September 1903):40.

21. *EMJ* 78 (1904):194, 233, 480, 486, 521, 526, 566, 721, 804, 807, 841; 79 (1905):302, 453, 1021; 80 (1905):367, 416, 420.

22. *EMJ* 89 (1910):186, 884, 1080; 90 (1901):526, 1177, 1272; 92 (November 1911):1005; 94 (October 1912):757.

23. *EMJ* 69 (1900) to 92 (1911), passim.

24. Certificate of incorporation, New Monarch Mining Company, CSA; *EMJ* 72 (October 1901):437, 504.

25. *EMJ* 72 (November 1901):644; 73 (1902):116, 149, 187, 268, 361, 424, 495, 563, 671, 806, 872; 74 (1902):125, 494, 827, 860; 75 (May 1903):722; Etienne A. Ritter, "The New Smelter at Salida," p. 813.

26. *EMJ* 74 (December 1902):827; 75 (1903):96–97, 130, 645, 722, 758; 76 (1903):63, 102, 138; certificate of incorporation, Republic Smelting and Refining Company, CSA.

27. *EMJ* 75 (1903):422, 645, 978; 76 (1903):102, 635, 751, 670, 1019.

28. *EMJ* 84 (February 1907):305.

29. *EMJ* 81 (1906) through 112 (1921), passim.

30. Henderson, *Mining in Colorado*, pp. 15–17; American Smelting and Refining Company, *Annual Report, 1961*.

Bibliography

A. Manuscript and Archival Material

Charles Francis Adams, Jr., Papers. Massachusetts Historical Society, Boston.

Adelberg and Raymond Collection. New York Public Library, New York, New York.

Atchison, Topeka, and Santa Fe Railroad Collection. Kansas State Historical Society, Topeka.

George F. Becker Papers. Library of Congress, Washington, D.C.

William A. Bell Papers, Colorado Historical Society, Denver.

John Jelling Blow Papers. University of Colorado, Boulder.

Rosemary Bogy Collection. Missouri Historical Society, Saint Louis.

David R. C. Brown Papers. University of Colorado, Boulder.

Brown University Archives. Brown University, Providence, Rhode Island.

George Jarvis Brush Collection. Yale University, New Haven, Connecticut.

John F. Campion Papers. University of Colorado, Boulder.

Colorado State Archives. Denver.

Commonwealth of Massachusetts Archives. Office of the Secretary of State, Boston.

James W. and E. Costello Converse Papers. Harvard University Graduate School of Business Administration, Boston, Massachusetts.

James V. Dexter Collection. Colorado Historical Society, Denver.

R. G. Dun & Company Records. Harvard University Graduate School of Business Administration, Boston, Massachusetts.

Thomas Egleston Papers. Columbia University, New York, New York.

Samuel F. Emmons Collection. Library of Congress, Washington, D. C.

James D. Hague Papers. Henry E. Huntington Library and Art Gallery, San Marino, California.

James D. Hague Papers. National Resources Manuscript Group. Yale University, New Haven, Connecticut.

John Hays Hammond Papers. Yale University, New Haven, Connecticut.

Harrison Family Collection. Missouri Historical Society, St. Louis.

Edwin Harrison Papers. Colorado Historical Society, Denver.

Abram S. Hewitt Papers. New York Historical Society, New York.

Henry Lee Higginson Collection. Harvard University Graduate School of Business Administration, Boston, Massachusetts.

Hill Collection. Denver Public Library.

Crawford and Louise Sneed Hill Collection. Colorado Historical Society, Denver.

Nathaniel P. Hill Collection. Colorado Historical Society, Denver.

George Hubbard Holt Collection. University of Colorado, Boulder.

Henry M. Howe Collection. Columbia University, New York, New York.

Nathaniel S. Keith Collection. Colorado Historical Society, Denver.

Clarence King Papers. Henry E. Huntington Library and Art Gallery, San Marino, California.

Nathaniel P. Langford Collection. Minnesota Historical Society, St. Paul (microfilm 1023 in Norlin Library, University of Colorado, Boulder).

James E. Lyon Record Book and Journal. Colorado Historical Society, Denver.

National Archives and Records Service.

Records of the Department of Justice, Record Group 60, Washington, D.C.

Records of the Department of Commerce, Record Group 122, Washington, D.C.

Records of the United States Circuit Court, Federal Records Center, Region 8, Denver, Colorado.

A. R. Meyer & Company Ore Milling and Sampling Company vs. J. B. Dixon. Docket 374.

Argentine Mining Company of St. Louis of the State of

Missouri vs. the Adelaide Consolidated Mining and Smelting Company. Docket 214.

Arkansas Valley Smelting Company vs. Belden Mining Company. Docket 1152.

Francis E. Everett and the Omaha Smelting and Refining Company vs. the Hukill Gold and Silver Mining Company. Docket 213.

James B. Grant vs. Council Bluffs Iron Works. Docket 417.

St. Louis Smelting and Refining Company vs. H. P. Parlon and D. Shaw. Docket 292.

Thomas C. McGowan vs. the La Plata Mining and Smelting Company. Docket 667.

United States vs. Nathaniel P. Hill and the Boston and Colorado Smelting Company. Docket 121.

Winfield Scott Keyes and Albert Arents vs. James B. Grant and James Grant. Docket 613.

Bryan Obear Scrapbooks. University of Missouri, Columbia.

Eben Olcott Papers. New York Historical Society, New York, New York.

Harper M. Orahood Papers. University of Colorado, Boulder.

George Foster Peabody Collection. Library of Congress, Washington, D.C.

Rossiter W. Raymond Papers. In the possession of the author.

William H. Reynolds Collection. Rhode Island Historical Society, Providence, Rhode Island.

John D. Rockefeller Papers. Rockefeller Archive Center, Pocantico Hills, New York.

Frederick Roeser Collection. Henry E. Huntington Library and Art Gallery, San Marino, California.

Henry Seligman Collection. Harvard University Graduate School of Business Administration, Boston, Massachusetts.

Eben Smith Correspondence. Denver Public Library.

Horace A. W. Tabor Collection. Colorado Historical Society, Denver.

Henry M. Teller Collection. Colorado Historical Society, Denver.

Henry M. Teller Collection. Denver Public Library.

Robert F. Weitbrec Collection. Colorado Historical Society, Denver.

William C. Whitney Collection. Library of Congress, Washington, D.C.

D. A. Wightman Collection. Harvard University Graduate School of Business Administration, Boston, Massachusetts.

Henry E. Wood Collection. Henry E. Huntington Library and Art Gallery, San Marino, California.

B. Public Documents

Federal

Bastin, Edson S. and Hill, James M. *Economic Geology of Gilpin County and Adjacent Parts of Clear Creek and Boulder Counties, Colorado*. U.S. Geological Survey, Professional Paper no. 94. Washington, D.C.: Government Printing Office, 1927.

Browne, J. Ross, and Taylor, James W. *Reports upon the Mineral Resources of the United States*. U.S. Treasury Department. Washington, D.C.: Government Printing Office, 1867.

Emmons, Samuel F. *Geology and Mining Industry of Leadville, Colorado*. U.S. Geological Survey, Monograph 12. Washington, D.C.: Government Printing Office. 1886.

Hague, James D. *Mining Industry*. Vol. 3 of *Report of the Geological Exploration of the Fortieth Parallel*, by Clarence King. Washington, D.C.: Government Printing Office, 1870.

Henderson, Charles W. *Mining in Colorado: A History of Discovery, Development and Production*. U.S. Geological Survey, Professional Paper no. 138. Washington, D.C.: Government Printing Office, 1926.

Raymond, Rossiter W. *Statistics of Mines and Mining in the States and Territories West of the Rocky Mountains*. 8 vols. Washington, D.C.: Government Printing Office, 1868–75. (Titles vary.)

U.S. Congress. House. *Report of the Industrial Commission on the Relations and Conditions between Capital and Labor Engaged in the Mining Industry, Including Testimony, Review of Evidence, and Topical Digest*. House Doc. 181, serial set 4342, 57th Cong., 1st sess. vol. 12, 1901.

U.S. Congress. House. *Report of the Industrial Commission on Trusts and Industrial Combinations*. House Doc. 182, serial set 4343, 57th Cong., 1st sess., vol. 13, 1901.

U.S. Congress. House. Committee on Ways and Means. *Revision of the Tariff: Hearings before the Committee on Ways and Means, 1889–90*. H. Misc. Doc. 176, serial set 2774, 51st Cong., 1st sess., 1889.

U.S. Department of the Interior. Geological Survey. *Mineral Resources of the United States, Calendar Year 1885*. Washington, D.C.: Government Printing Office, 1886.

U.S. Department of the Treasury. *Report of the Director of the Mint*

upon *the Statistics of the Production of Precious Metals in the United
States*. 4 vols. Washington, D.C. Government Printing Office,
1881–84. (Titles vary.)

State

Colorado Bureau of Labor Statistics. *Seventh Biennial Report, 1899–
1900*. Denver: Smith-Brooks Printing Company, 1900.
Colorado General Assembly. House. *House Journal of the General
Assembly of the State of Colorado, Second Session*. Denver: Times
Steam Printing House, 1879.
———. *House Journal of the General Assembly of the State of Colorado,
Fifth Session*. Denver: Collier and Cleveland Lithograph Company,
1886.
Colorado General Assembly. Senate. *Senate Journal of the General
Assembly of the State of Colorado, Second Session*. Denver: Times
Steam Printing House, 1879.
———. *Senate Journal of the General Assembly of the State of Colorado,
Fifth Session*. Denver: Collier and Cleveland Lithograph Company,
1886.
Colorado Legislative Assembly. Council. *Council Journal of the Legis-
lative Assembly of the Territory of Colorado, Ninth Session*. Central
City: D. C. Collier, Register Office, 1872.

C. Books and Pamphlets

American Smelting and Refining Company. *Annual Reports*. 1900–
1961.
Athearn, Robert G. *Rebel of the Rockies: A History of the Denver and
Rio Grande Western Railroad*. New Haven: Yale University Press,
1962.
———. *Union Pacific Country*. Chicago: Rand McNally, 1971.
Barclay, R. E. *Ducktown Back in Raht's Time*. Chapel Hill: University
of North Carolina Press, 1946.
Baskin and Company, O. L. *History of the City of Denver, Arapahoe
County, and Colorado*. Chicago, 1880.
Bennett, Charles W.; Van Horne, W. G.; and Carter, Thomas H. *Before
the Hon. Secretary of the Treasury. In the Matter of the Hearing on the
Question of Admitting Free of Duty Foreign Ores Containing both
Lead and Silver. Argument and Authorities*. Washington, D.C.: Judd
and Detweiler, 1889.

Bernstein, Marvin D. *The Mexican Mining Industry, 1890–1950: A Case in the Interaction of Politics, Economics, and Technology.* Albany: State University of New York, 1964.

Blackmore, William. *Colorado: Its Resources, Parks, and Prospects as a New Field for Emigration.* . . . London: Sampson Low, Son, and Marston, 1869.

By-laws of the Colorado Reduction Works. N.p., n.d.

Catalogue of the Officers and Students of Brown University, 1854–1855 [through] 1864–1865. N.p., n.d.

Committee on Stock List, New York Stock Exchange. *Consolidated Kansas City Smelting and Refining Company.* New York, November 1, 1898.

Committee on Unlisted Securities, New York Stock Exchange. *American Smelting and Refining Company.* New York, April 20, 1899.

Cushman, Samuel, and Waterman, J. P. *The Gold Mines of Gilpin County, Historical, Descriptive, and Statistical.* Central City: Register Steam Printing House, 1876.

Dawson, Thomas Fulton. *Life and Character of Edward O. Wolcott Late a Senator of the United States from the State of Colorado.* New York: Knickerbocker Press, 1911.

Examination by Chemical Analysis and Otherwise, of Substances Emptied into the Public Waters of the State, from Gas and Other Manufactories, Sewerage and Other Sources. . . . Providence, 1861.

F. Anton Eilers. N.p., n.d.

Fahey, John. *The Ballyhoo Bonanza: Charles Sweeny and the Idaho Mines.* Seattle: University of Washington Press, 1971.

———. *D. C. Corbin and Spokane.* Seattle: University of Washington Press, 1965.

First Annual Report of the President and Directors of the Brown Silver Mining Company, of Colorado, for the Year Ending March 2, 1868. Philadelphia: James B. Rodgers, 1868.

Fossett, Frank. *Colorado. Its Gold and Silver Mines, Farms and Stock Ranges, and Health and Pleasure Resorts. Tourist's Guide to the Rocky Mountains.* New York: C. G. Crawford, 1879.

Gates, William B., Jr. *Michigan Copper and Boston Dollars: An Economic History of the Michigan Copper Mining Industry.* Cambridge: Harvard University Press, 1951.

Griswold, Don, and Griswold, Jean. *Carbonate Camp Called Leadville.* Denver: University of Denver Press, 1951.

Grodinsky, Julius. *Transcontinental Railway Strategy, 1869–1893: A*

Study of Businessmen. Philadelphia: University of Pennsylvania Press, 1962.

Hill, Nathaniel P. *Speeches and Papers on the Silver, Postal Telegraph, and Other Economic Questions*. Colorado Springs: Gazette Publishing Company, 1890.

————. *Use of Silver as a Money Metal*. N.p., n.d.

Historical Statistics of Brown University, 1764–1934. Providence: Brown University, 1936.

Hixon, Hiram W. *Notes on Lead and Copper Smelting and Converting*. 4th ed. New York: Hill Publishing Company, 1914.

Hofman, H. O. *Metallurgy of Copper*. New York: McGraw-Hill, 1914.

————. *The Metallurgy of Lead and the Desilverization of Base Bullion*. 6th ed. New York: Scientific Publishing Company, 1901.

Hollister, Ovando, J. *The Mines of Colorado*. Springfield, Mass.: Samuel Bowles, 1867.

Hoyt, Edwin P., Jr. *The Guggenheims and the American Dream*. New York: Funk and Wagnalls, 1967.

The Industries of St. Louis. Her Relation as a Center of Trade. Manufacturing Establishments and Business Houses. St. Louis: J. M. Elstner and Company, 1879.

Ingalls, Walter Renton. *Lead and Zinc in the United States, Comprising an Economic History of the Mining and Smelting of the Metals and the Conditions Which Have Affected the Development of the Industries*. New York: Hill Publishing Company, 1908.

Ingersoll, Ernest. *The Crest of the Continent: A Record of a Summer's Ramble in the Rocky Mountains and Beyond*. Chicago: R. R. Donnelley and Sons, 1885.

Karnes, Thomas L. *William Gilpin, Western Nationalist*. Austin: University of Texas Press, 1970.

Kent, L. A. *Leadville, The City. Mines and Bullion Product. Personal Histories of Prominent Citizens. Facts and Figures Never Before Given to the Public*. Denver: Daily Times Steam Printing Company and Blank Book Manufactury, 1880.

King, Joseph E. *A Mine to Make a Mine: Financing the Colorado Mining Industry, 1859–1902*. College Station: Texas A & M University Press, 1977.

Marcosson, Isaac F. *Metal Magic: The Story of the American Smelting and Refining Company*. New York: Farrar, Straus, 1949.

Marriner, Ernest Cummings. *History of Colby College*. Waterville, Maine: Colby College Press, 1963.

Mathews, Alfred E. *Pencil Sketches of Colorado, Its Cities, Principal Towns and Mountain Scenery*. New York: J. Bien, 1866.

Minchinton, W. E., ed. *Industrial South Wales, 1750–1914: Essays in Welsh Economic History*. New York: Augustus M. Kelley Publishers, 1969.

National Cyclopaedia of American Biography. New York: James T. White, 1906.

O'Connor, Harvey. *The Guggenheims: The Making of an American Dynasty*. New York: Covici Friede, 1937.

Paul, Rodman Wilson. *Mining Frontiers of the Far West, 1848–1880*. New York: Holt, Rinehart and Winston, 1963.

Percy, John. *Metallurgy: The Art of Extracting Metals from Their Ores*. Part 1. *Silver and Gold*. London: John Murray, 1880.

――――. *The Metallurgy of Lead, Including Desilverization and Cupellation*. London: John Murray, 1870.

Peters, Edward D., Jr. *Modern Copper Smelting*. 7th ed. New York: Scientific Publishing Company, 1895.

――――. *The Principles of Copper Smelting*. New York: Hill Publishing Company, 1907.

Peters, Eleanor Bradley. *Edward Dyer Peters (1849–1917)*. New York: Knickerbocker Press, 1918.

Petrowski, William Robinson. *The Kansas Pacific: A Study in Railroad Promotion*. Ann Arbor, Mich.: University Microfilms, 1967.

Poor, M. C. *Denver South Park & Pacific: A History of the Denver South Park & Pacific Railroad and Allied Narrow Gauge Lines of the Colorado & Southern Railway*. Denver: World Press, 1949.

Prospectus of the Belle Monte Furnace Iron and Coal Company of Boulder County, Colorado. N.p, n.d.

Prospectus. *Lake City Mining and Smelting Company*. N.p., n.d.

――――. *La Plata Mining and Smelting Company*. N.p., n.d.

――――. *The Pioneer Smelting Company, of Colorado*. N.p., n.d.

The Pueblo & Oro Railroad. New York: E. Wells Sackett and Brother, 1878.

Rickard, Thomas A., ed. *Rossiter Worthington Raymond: A Memorial*. New York: American Institute of Mining Engineers, 1920.

San Juan Smelting and Refining Company. *Annual Reports*. 1887–93.

Scamehorn, H. Lee. *Pioneer Steelmaker in the West: The Colorado Fuel and Iron Company, 1872–1903*. Boulder, Colo.: Pruett Publishing Company, 1976.

Sheedy, Dennis. *The Autobiography of Dennis Sheedy*. N.p., n.d.

Smiley, Jerome C. *History of the City of Denver with Outlines of the*

Earlier History of the Rocky Mountain Country. Denver: J. H. Williamson, 1903.

Smith, Duane A. *Horace Tabor: His Life and the Legend*. Boulder: Colorado Associated University Press, 1973.

Southern Colorado Coal and Town Company. *First Annual Report, 1878*. Colorado Springs: Gazette Publishing Company, 1879.

Spence, Clark C. *British Investments and the American Mining Frontier: 1860–1901*. Ithaca: Cornell University Press, 1958.

————. *Mining Engineers and the American West: The Lace-Boot Brigade, 1849–1933*. New Haven: Yale University Press, 1970.

Stockholders Committee of Investigation. *Report to the Stockholders of the American Smelting and Refining Company*. New York: Evening Post Job Printing Office, 1922.

Taylor, Bayard. *Colorado: A Summer Trip*. New York: G. P. Putnam and Son, 1867.

Tice, John H. *Over the Plains and on the Mountains; or Kansas and Colorado, Agriculturally, Mineralogically, and Aesthetically Described*. St. Louis: "Industrial Age" Printing Company, 1872.

Wagner, Henry Raup. *Bullion to Books*. Los Angeles: Zamorano Club, 1942.

Whitney, Joel Parker. *Silver Mining Regions of Colorado, with Some Account of the Different Processes Now Being Introduced for Working the Gold Ores of That Territory*. New York: D. Van Nostrand, 1865.

Williams, Gatenby [William Guggenheim], in collaboration with Heath, Charles Monroe. *William Guggenheim*. New York: Lone Voice Publishing Company, 1934.

D. Periodicals

Engineering and Mining Journal. Vols. 17–114. New York: Scientific Publishing Company, 1874–1922.

Mining and Scientific Press. Vols. 19–68. San Francisco: Dewey Publishing Company, 1869–94.

Mining Journal. Vol. 39. London, 1869.

Transactions of the Rhode Island Society for the Encouragement of Domestic Industry in the Year 1859.

E. Articles

Collins, George E. "Colorado." *Engineering and Mining Journal* 89 (January 1910):97–99.

Courtis, William M. "The Wyandotte Silver Smelting and Refining Works." *Transactions of the American Institute of Mining Engineers* 2 (1873–74):89–101.

Dwight, Arthur S. "A Brief History of Blast-Furnace Lead Smelting in America." *Transactions of the American Institute of Mining Engineers* 121 (1936):9–37.

Egleston, Thomas. "Boston and Colorado Smelting Works." *Transactions of the American Institute of Mining Engineers* 4 (1876): 276–98.

Eilers, Anton. "Coke from Lignites." *Transactions of the American Institute of Mining Engineers* 2 (1874):101–2.

Frazer, Persifor. "Colorado Notes." *Engineering and Mining Journal* 30 (August 1880):107–8, 139–40.

Frost, O. J. "The Grant Lead-Smelting Works." *Engineering and Mining Journal* 35 (March 1883):163–64.

Furman, H. Van F. "Metallurgy of Copper." *Mines and Minerals* 20 (August 1899):25–27; 20 (September 1899):79–81.

Glazer, Sidney, ed. "A Michigan Correspondent in Colorado, 1878." *Colorado Magazine* 37 (July 1960):207–18.

Guiterman, Franklin. "Status of Mining and Smelting in Colorado." *Engineering and Mining Journal* 90 (November 1910):1109–10.

Guiterman, Kenneth S. "Mining Coal in Southern Colorado." *Engineering and Mining Journal* 88 (November 1909):1009–15.

Hahn, Otto H. "A Campaign in Railroad District, Nevada." *Transactions of the American Institute of Mining Engineers* 3 (1874–75): 329–32.

Hahn, Otto H.; Eilers, Anton; and Raymond, Rossiter W. "The Smelting of Argentiferous Lead Ores in Nevada, Utah, and Montana." *Transactions of the American Institute of Mining Engineers* 1 (1871–73):91–131.

Hale, Jesse D. "The First Successful Smelter in Colorado." *Colorado Magazine* 13 (September 1936):161–67.

Hoyt, A. W. "Over the Plains to Colorado." *Harper's New Monthly Magazine* 35 (June 1867):1–21.

Ingalls, Walter Renton. "Franz Fohr." *Engineering and Mining Journal* 108 (November 1919):828–9.

Jernegan, Joseph L. "Lead and Silver Smelting in Chicago." *Transactions of the American Institute of Mining Engineers* 2 (1873–74):279–95.

———. "Notes on a Metallurgical Campaign at Hall Valley, Colorado." *Transactions of the American Institute of Mining Engineers* 5 (1876–77):560–75.

———. "The Swansea Silver Smelting and Refining Works of Chicago." *Transactions of the American Institute of Mining Engineers* 4 (1875–76):35–53.

———. "The Whale Lode of Park County, Colorado Territory." *Transactions of the American Institute of Mining Engineers* 3 (1874–75):352–56.

Kedro, M. James. "Czechs and Slovaks in Colorado, 1860–1920." *Colorado Magazine* 54 (spring 1977):93–127.

MacFarlane, Thomas. "On the Use of Determining Slag Densities in Smelting." *Transactions of the American Institute of Mining Engineers* 8 (1879–80):71–74.

———. "Silver Islet." *Transactions of the American Institute of Mining Engineers* 8 (1879–80):226–53.

" 'Mexico' or 'Mexican Town': Picturesque Settlement of Smelter Laborers in the Heart of Pueblo." *Camp and Plant* 4 (September 1903):245–50.

"Otto H. Hahn." *Engineering and Mining Journal* 100 (August 1915):362.

Pearce, Harold V. "Improved Method of Slag-Treatment at Argo." *Transactions of the American Institute of Mining Engineers* 39 (1908):89–100.

———. "The Pearce Gold-Separation Process." *Transactions of the American Institute of Mining Engineers* 39 (1908):722–34.

Pearce, Richard, "Progress of Metallurgical Science in the West." *Transactions of the American Institute of Mining Engineers* 17 (1889):55–72.

"The Pearce Turret Furnace." *Engineering and Mining Journal* 55 (June 1893):513.

Peters, Edward D., Jr. "The Mount Lincoln Smelting Works, at Dudley, Colorado." *Transactions of the American Institute of Mining Engineers* 2 (1874):310–14.

———. "Reverberatory Practice at Argo, Colorado." *Engineering and Mining Journal* 50 (August 1890):189–90.

Pufahl, O. "The Globe Plant of the American Smelting and Refining Company." *Engineering and Mining Journal* 88 (November 1909):1009–15.

———. "The Pueblo Lead Smelters." *Engineering and Mining Journal* 81 (May 1906):889.

———. "The Works of the United States Zinc Company at Pueblo, Colorado." *Engineering and Mining Journal* 81 (June 1906):1232.

Raymond, Rossiter W. "Anton Eilers." *Engineering and Mining Journal* 103 (April 1917):762–64.

———. "Henry Williams." *Engineering and Mining Journal* 74 (July 1902):82.

———. "Remarks on the Precipitation of Gold in a Reverberatory Hearth." *Transactions of the American Institute of Mining Engineers* 1 (1871–73):320–22.

Rickard, T. A. "Richard Pearce: The Biographic Sketch of a Pioneer Metallurgist." *Engineering and Mining Journal* 125 (March 1928):404–9.

Ritter, Etienne A. "The New Smelter at Salida." *Engineering and Mining Journal* 74 (December 1902):813.

Steele, B. W. "Hon. Nathaniel P. Hill, Ex-United States Senator of Colorado." *National Magazine* 15 (February 1892):431–33.

Tonge, Thomas. "Smelting Gold and Silver Ores in Colorado. . . ." *Mines and Minerals* 19 (October 1898):98–100.

Ulke, Titus. "Present Practice in Copper Concentration and Extraction." *Mineral Industry: Its Statistics, Technology and Trade for 1893* 2:87–98.

F. Newspapers

Colorado Daily Chieftan (Pueblo), 1878–80.

Colorado Miner (Georgetown), October 14, 1869.

Colorado Transcript (Golden), June–October 1867, and October 6, 1869.

Daily Colorado Tribune (Denver), July 29, 1868.

Daily Miner's Register (Black Hawk), June 18, 19, 1868.

Daily Mining Journal (Black Hawk), August 29, 1866.

Denver Daily Times, January 1, 1878–January 1, 1879.

Denver Republican, January 1, 1883–January 1, 1910.

Denver Tribune, January 1, 1879.

Globe-Democrat (Saint Louis), October 16, 1898.

Pueblo Daily Chieftan (Pueblo), 1881–85.

Republican (Saint Louis), January 1, 1874.

Rocky Mountain News (Denver), July 1, 1866–January 1, 1910.

G. Theses and Dissertations

Ambrosius, Carl E. "The Method of Smelting and Refining Silver at the Boston and Colorado Smelter." Undergraduate thesis, Colorado School of Mines, 1888.

Hakola, John W. "Samuel T. Hauser and the Economic Development of Montana: A Case Study in Nineteenth-century Frontier Capitalism." Ph.D. diss., University of Indiana, 1961.

Lonsdale, David L. "The Movement for an Eight-hour Law in Colorado, 1893–1913." Ph.D. diss., University of Colorado, 1963.

McLatchy, Patrick Henry. "A Collection of Data on Foreign Corporate Activity in the Territory of Montana, 1864–1889." M.A. thesis, Montana State University, 1961.

Perry, Charles M. "Chemistry and Its Teachers in Brown University." Undergraduate thesis, Brown University Archives, about 1935.

Spence, Clark Christian. "Robert Orchard Old and the British and Colorado Mining Bureau." M.A. thesis, University of Colorado, 1951.

Toole, K. Ross. "A Study of the Anaconda Copper Mining Company: A Study in the Relationship between a State and Its People and a Corporation, 1880–1950." Ph.D. diss., University of California at Los Angeles, 1954.

Traxler, Ralph Newton, Jr. "Some Phases of the History of the Colorado Central Railroad, 1865–1885." M.A. thesis, University of Colorado, 1947.

H. Interviews

Beer, Andrew E., New York, New York, July 11, 1974.

Bennett, John. 1884. Bancroft Collection, University of California at Berkeley. Copy in Norlin Library, University of Colorado, Boulder, Colorado.

Fohr, Franz. Leadville, Colorado, 1884. Bancroft Collection, University of California at Berkeley. Copy in Norlin Library, University of Colorado, Boulder, Colorado.

Pope, Darwin J. and Pope, Mary. New York, New York, July 11, 1974.

Potter R. B., May 24, 1886. Bancroft Collection, University of California at Berkeley. Copy in Norlin Library, University of Colorado, Boulder, Colorado.

Index

325